Emerging Trends in Mechatronics

Series Editor

Aydin Azizi, Oxford, UK

Aydin Azizi · Reza Vatankhah Barenji
Editors

Industry 4.0

Technologies, Applications, and Challenges

 Springer

Editors
Aydin Azizi
School of Engineering
Computing and Mathematics
Oxford Brookes University
Oxford, UK

Reza Vatankhah Barenji
Department of Engineering
School of Science and Technology
Nottingham Trent University
Nottingham, UK

ISSN 2731-4855   ISSN 2731-4863 (electronic)
Emerging Trends in Mechatronics
ISBN 978-981-19-2014-1   ISBN 978-981-19-2012-7 (eBook)
https://doi.org/10.1007/978-981-19-2012-7

This Springer imprint is published by the registered company Springer Nature Singapore Pte Ltd.
The registered company address is: 152 Beach Road, #21-01/04 Gateway East, Singapore 189721,
Singapore

Contents

Industry 4.0 Concepts, Technologies, and Its Ecosystem

Kubra Nur Ozcan, Ozge Yesilyurt, Serap Demir, and Batihan Konuk

1 Introduction

Simply expressed, industry refers to a country's production of commodities and services. The first, second, and third industrial revolutions, respectively, are ex-post referred to as the successive stages of industrialization [1].

Industry 4.0, also known as "Smart Manufacturing" or "Fourth Industrial Revolution," was first conceived in Germany in 2011 during the Hannover Fair by German academics, practitioners, and government officials. Industry 4.0 and Smart factory are both terms that are used interchangeably when referring to smart manufacturing. Some refer to it as the "Industrial Internet of Things." Given the massive data volumes created by the system on a daily basis, smart manufacturing is the convergence and fusion of innovative digital technologies that are projected to change the dimensions of the manufacturing sector beyond conception [2].

The digitalization of the manufacturing industry is a hallmark of the Fourth Industrial Revolution, as is the explosion in data volume, processing power, and connection (especially in new low power networks). Industry 4.0 refers to a digital revolution that is currently sweeping the world, with the goal of digitizing the whole manufacturing process with minimal human or physical interaction. The goal is to cover as many industries as possible while adapting and improving existing technology to meet the demands of digital manufacturing. High-performance information technology offers new options for analytics and corporate intelligence, as well as unconventional human–machine communication, such as touch interfaces and the like [3].

K. N. Ozcan (✉) · O. Yesilyurt · S. Demir · B. Konuk
Department of Industrial Engineering, Hacettepe University, Beytepe Campus 06800, Ankara, Turkey
e-mail: kkubranur.demir@gmail.com

© The Author(s), under exclusive license to Springer Nature Singapore Pte Ltd. 2023
A. Azizi and R. V. Barenji (eds.), *Industry 4.0*, Emerging Trends in Mechatronics,
https://doi.org/10.1007/978-981-19-2012-7_1

Smart manufacturing, smart factories, and the Industrial Internet of Things (IIoT) are just a few of the Industry 4.0 buzzwords. Global rivalry is fueled by the ever-changing expectations of today's consumers. These expectations necessitate a significant shift in the manufacturing process. The shift and ever-increasing client needs and expectations for personalized products, faster-than-ever replies, and companies' responses in meeting these demands and remaining competitive are important drivers of this revolution. Industry 4.0 is a bright new light of hope for achieving this goal. It brings together business, manufacturing, suppliers, and customers [1]. Phases and key contributions of industrial revolution are indicated in the Fig.1.

1.1   The First Industrial Revolution (1765)

The bulk of people in the pre-industrial era lived in small, rural settlements, reliant on agriculture for survival and money, with low wages and widespread malnutrition and disease. The majority of people's food and clothing needs were met by individuals, and the tiny amount of manufacturing that was done was done at home or in small shops using rudimentary hand tools and machines [1].

The first industrial revolution ushered in a pivotal period in human history, affecting practically every area of daily life. From 1760 through 1840, production settings, particularly industry, were destined to undergo significant changes. Great Britain was the first country to transition from an agricultural to an industrial economy, followed by numerous European countries such as Belgium, France, and Germany [4].

Textile industries were the largest industries in terms of capital investment, output, and number of jobs offered during this time period, and they were also the first to apply contemporary manufacturing processes. The introduction of coal, iron, railroads, and textiles were all major advancements in the first industrial revolution, which began in 1764 with James Hargreaves' discovery of the spinning jenny. Because of the growth of railroads and the acceleration of economic, human, and material transformations, mass coal extraction and the advent of the steam engine provided a new type of energy that propelled all processes forward [5]. Figure 2 illustrates the first mechanical loom.

1.2   The Second Industrial Revolution (1870)

Despite some parallels to the first industrial revolution, the second industrial revolution, sometimes referred to as the technological revolution, took place between 1870 and 1914. This brief period of history saw a renewed and stronger acceleration of the industrialization process, owing to scientific breakthroughs such as the development of lighter materials such as alloys and synthetics, as well as the adoption of a new energy source, namely electricity. Within manufacturing facilities, the first assembly

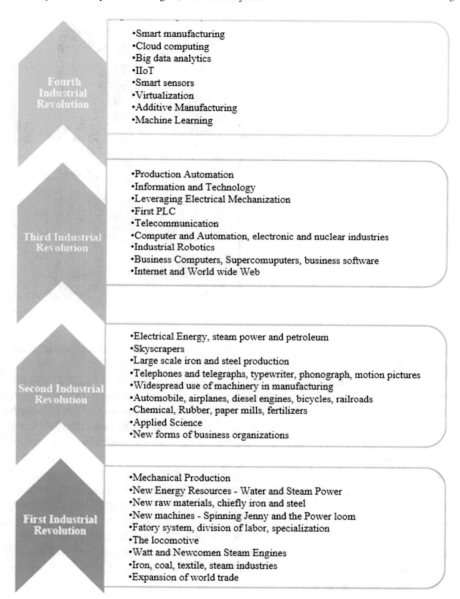

Fourth Industrial Revolution
- Smart manufacturing
- Cloud computing
- Big data analytics
- IIoT
- Smart sensors
- Virtualization
- Additive Manufacturing
- Machine Learning

Third Industrial Revolution
- Production Automation
- Information and Technology
- Leveraging Electrical Mechanization
- First PLC
- Telecommunication
- Computer and Automation, electronic and nuclear industries
- Industrial Robotics
- Business Computers, Supercomuputers, business software
- Internet and World wide Web

Second Industrial Revolution
- Electrical Energy, steam power and petroleum
- Skyscrapers
- Large scale iron and steel production
- Telephones and telegraphs, typewriter, phonograph, motion pictures
- Widespread use of machinery in manufacturing
- Automobile, airplanes, diesel engines, bicycles, railroads
- Chemical, Rubber, paper mills, fertilizers
- Applied Science
- New forms of business organizations

First Industrial Revolution
- Mechanical Production
- New Energy Resources - Water and Steam Power
- New raw materials, chiefly iron and steel
- New machines - Spinning Jenny and the Power loom
- Fatory system, division of labor, specialization
- The locomotive
- Watt and Newcomen Steam Engines
- Iron, coal, textile, steam industries
- Expansion of world trade

Fig. 1 Phases and key contributions of industrial revolution [1]

Fig. 2 First mechanical loom Doncaster England, 1784

lines were constructed, laying the groundwork for the future phenomenon of mass production [4].

In addition, the steel industry began to develop and flourish in tandem with the ever-increasing demand for steel. With the creation of the telegraph and telephone, as well as the introduction of the vehicle and plane around the turn of the twentieth century, communication and transportation systems were transformed. Disinfectants and antiseptics, including phenol and bromines, the role of bacteria in wound infection, salicylic acid, and other artificial materials were also developed by chemistry. During this revolution, electric generators, vacuum pumps, gas lighting systems, and transformers were invented. Electricity was identified as a universal energy transmission system. Railways have become faster, diesel engines have been invented, and clipper ships have been developed in the field of autos [5].

On the other hand, the second industrial revolution resulted in widespread poverty, job loss as humans were replaced by machines, deep depression, and widespread economic insecurity. With the outbreak of World War I, the second industrial revolution came to an end [1].

1.3 The Third Industrial Revolution (1969)

Partially automated processes employing memory-programmable controllers and computers kicked off the Third Industrial Revolution in the 1970s. Following the

Fig. 3 Programmable logic
controller (PLC), 1969

implementation of these innovations, we are now able to optimize the entire development process without the need for human intervention. The move from analogue and mechanical systems to digital systems, often known as the Digital Revolution, has occurred [6].

With this revolution came the introduction of electronics and computers, with the goal of automating production. Telecommunication was made available. In 1870, the first Programmable Logic Controller (PLC) was released. PLC is illustrared in Fig. 3. Factory production was controlled by PLCs. Robots were created. This revolution surpassed its two prior revolutions due to the introduction and utilization of nuclear energy [5].

This new technology paved the way for the development of tiny materials, which would open doors to a variety of fields, including space exploration and biotechnology. During the third revolution, electronics and information technology were used to automate production [5].

1.4 The Fourth Industrial Revolution

In the twenty-first century, the fourth industrial revolution, dubbed "Industry 4.0," is currently underway. It is based on the concept of smart manufacturing. The objective is to use digitization to create, construct, and develop a robust virtual environment capable of influencing the physical world. The objective is to connect all manufacturing lines and enable real-time communication between them [1].

Which places a higher premium on real-time data processing and heavily relies on interconnectivity via IoT and machine learning. Industry 4.0 connects manufacturing systems to the Internet of Things and the industrial Internet, enabling machines to communicate with one another and make intelligent decisions based on the system's algorithm. Industry 4.0 encompasses artificial intelligence, automated robots, flexible manufacturing automation systems, additive manufacturing, and augmented reality [5].

Although Germany pioneered Industry 4.0, similar principles have been implemented in countries worldwide. General Electric introduced the "Industrial Internet" to North America in late 2012 as a tight integration of the physical and virtual worlds based on IoT and big data analytics. This concept is applicable to a broad range of industries, including energy distribution and generation, health care, transportation, a variety of government sectors, manufacturing, and mining [1].

Numerous occupations have been transformed as a result of Industry 4.0. People have always been required to learn new daily chores, but they are now also required to use high-tech devices, which are rapidly gaining prominence in their professional lives. Industry 4.0 is defined as the digitalization and automation of the entire enterprise, including the manufacturing process [4]. The important innovations in the industrial revolution from past to present are shown in Fig. 4.

After examining each of the IRs, the authors come to the conclusion that smart design engineering should be able to deal with the benefits and obstacles presented by each perspective. It is summarized in Fig. 5.

- Automation gave way to autonomy and communication as technology progressed.
- Production: The factory's efficiency was boosted by automating simple and repetitive procedures and moving on to more complicated activities that require autonomous decision-making based on distributed data.
- The customer's purchasing decision is influenced not just by product cost and availability, but also by how well the product satisfies his or her specific requirements.
- The marketing function has grown to become more precise, allowing for more current information about particular clients.
- Product complexity and scope have evolved over time, and related services are now regarded PSSs as part of the supplied solution.
- A more continuous Product Development Process (PDP) is indicated by the design and development process, the increase of product complexity and scope, and, lastly, the ability to get feedback on the product while it is being used.
- Sustainability: The understanding of the factory's environmental impact, which was initially focused on industrial employees and then on pollution, has expanded to include numerous facets of sustainability, including the necessity for resource conservation and the formation of circular economies [7].

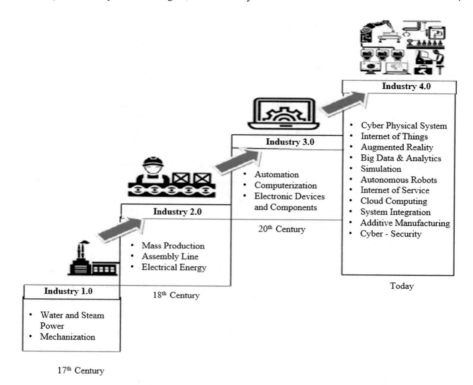

Fig. 4 The Timeline of the industrial revolutions [4]

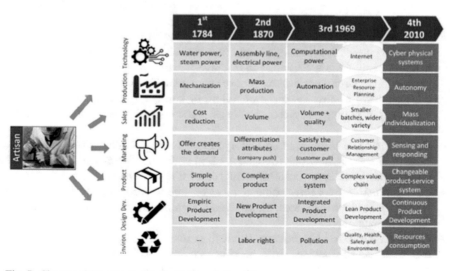

Fig. 5 Changes driven by each industrial revolution [7]

1.5   Industry 4.0 Design Principles

Manufacturers can use the design concepts to look into a possible transition to Industry 4.0 technology. The design principles are as follows:

- *Interoperability*: Interoperability refers to a system's components' capacity to collaborate within and across organizational boundaries in order to improve the delivery of services to individuals and communities. Interoperability is the ability of a product or system with well-understood interfaces to function with other products or systems, now or in the future, in terms of implementation or access, without limitations. While the phrase was originally used to refer to information technology or systems engineering services that facilitate information interchange, a more comprehensive definition considers social, political, and organizational aspects that influence system-to-system performance. As a result, when the separate components are technically diverse and maintained by different entities, interoperability entails the effort of creating cohesive services for users.
- *Virtualization*: CPSs must be able to construct and replicate a virtual replica of the real world. CPSs must also be able to monitor objects in their immediate surroundings. Simply said, everything must have a virtual duplicate. The term "virtualization" refers to the process of creating a virtual version of a device or system by dividing the resource into one or more execution contexts using a framework. The act of generating a virtual (rather than physical) version of something, such as virtual computer hardware platforms, storage devices, and computer network resources, is known as virtualization or virtualization in computing. Virtualization was first introduced in the 1960s as a way of conceptually partitioning the system resources given by mainframe computers among distinct applications. Since then, the term's meaning has evolved.
- *Decentralization*: Decentralization is the practice of distributing or delegating an organization's functions, particularly those related to planning and decision-making, away from a central, authoritative location or group. As a result, the production environment becomes more adaptable. The problem is outsourced to a higher level in circumstances of failure or conflicting goals. However, even with such technology in place, quality assurance is still required throughout the process.
- *Real-Time Capability*: A smart factory must be able to collect data in real time, store or analyze it, and make decisions based on fresh information. This includes not only market research but also internal procedures such as a machine breakdown on a production line. Smart objects must be able to recognize the problem and reassign jobs to other devices. This also adds significantly to production flexibility and optimization.
- *Customer-Oriented Production*: Production must be customer-focused. People and smart objects/devices must be able to connect effectively via the Internet of Services in order to produce products that meet the needs of customers. The Internet of Services becomes critical at this point.

- **Modularity**: In general, modularity refers to the ability of a system's components to be separated and recombined, providing flexibility and variety of application. Modularity is a notion that is used to minimize complexity by breaking down a system into variable degrees of interdependence and independence across components and "hide the complexity of each part behind an abstraction and interface." The concept of modularity, on the other hand, may be applied to a variety of fields, each with its own set of subtleties. A Smart Factory's ability to adapt to a new market is critical in a dynamic market. An average corporation would probably take a week to research the market and adjust its output accordingly in a typical instance. Smart factories, on the other hand, must be able to adjust quickly and smoothly to seasonal changes and market trends.

2 Technology Pillars of Industry 4.0

The six technologies of Industry 4.0, as indicated in Fig. 6, are the ones that this revolution incorporates. The following are the primary concepts and technologies involved in Industry 4.0.

2.1 *Internet of Things*

The Internet of Things (IoT) is a pervasive technology in today's society, and a Forbes report estimates that the number of IoT devices will reach 75.4 billion by 2025. The Internet of Things (IoT) devices connect the Internet to sensors via wired and wireless sensor networks (WSN), embedded systems, and radio-frequency identification (RFID) (RFID). Due to the integrated information system and automatic recovery of dynamic data, IoT sensors will revolutionize the industrial environment, from the shop floor to final customer delivery. An IoT system simplifies machine-to-machine communication, information exchange between components, dynamic data storage in the cloud, robot interaction, and item tracing throughout manufacturing, storage, and transportation. Dynamic data collection enables the development of effective decision support systems [8].

IoT business cases will be attractive in 2019 because of the combination of AI, machine learning, and contextually rich, real-time data streams given by IoT sensors and networks. The Internet of Things (IoT) is at the heart of many companies' digital revolutions, allowing them to improve existing operations while also developing and exploring innovative new business models as shown in Fig. 7 [9]. Cisco has forecasted that 500 billion things will be connected and linked to the Internet by 2030 [10].

The next sections go through the four aspects of the theoretical framework seen in Fig. 8.

Bottom left: The range of values that each construct in the theory could span must first be specified in the state's space that coincides with the theory's boundary.

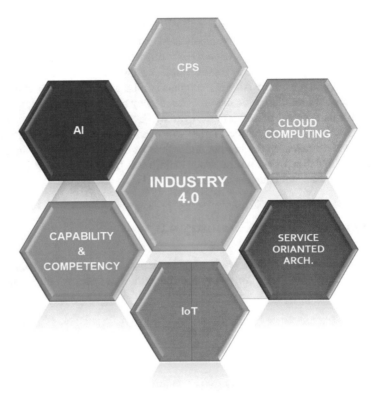

Fig. 6 Key technologies of industry 4.0

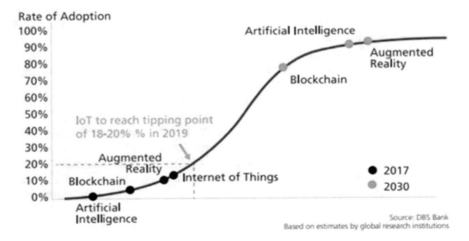

Fig. 7 IoT adoption to approach 100% over the next 10 years

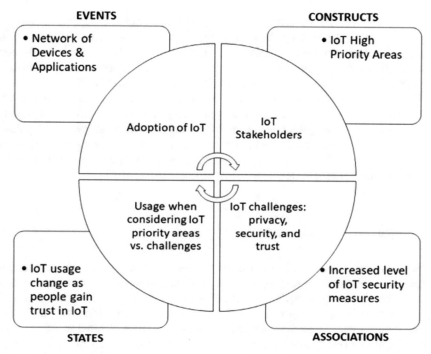

Fig. 8 IoT theoretical framework and conceptual model

When assessing IoT priority areas vs. problems, the left bottom quarter of the circle identified as the inside border state is Usage. Uncertainties, particularly privacy and security concerns, may have an influence on consumption. In certain circumstances, users have been ready to incur risks in this area if the rewards are significant enough. IoT use changes as individuals acquire faith in IoT, as illustrated by the outer border state in the model's left bottom rectangle [11]. As with other technologies, it is expected that as consumers acquire trust in the system, utilization will rise, which should be tested through future studies. "The events space that falls inside the theory's border must likewise be described [12]" says the left top.

Adoption of IoT is the inner border event in the left upper quadrant of the circle. This theoretical framework is centered on IoT in the workplace, which is intimately linked to the outside border events depicted in the model's left top rectangle— Network of Devices and Applications. Top right: IoT stakeholders are represented inside the right upper quarter of the circle, which is described as the inner boundary construct. In a theory, a construct represents an attribute of a class of entities in its domain in general [12]. He also underlines the need to carefully define the classes of objects to which qualities, in general, apply so that the meanings of each class and the items within each class are apparent. Otherwise, the nature of the objects covered by the theory will be unclear. Humans who are affected by or interested in the Internet of Things are referred to as stakeholders. IoT high priority regions are included in the outer boundary constructs indicated in the right top rectangle of the model. "A theory usually does not account for all the connections between its components. Rather, researchers aim to make informed judgments about which links to include in the theory and which to exclude. The absence of a construct-to-construct relationship in a theory does not always imply that none exists" [12].

The rapid growth of the Internet of Things is allowing numerous new applications to arise, as well as the redesign of old systems to make them more efficient. In reality, sensors are being implanted in an increasing number of things, allowing them to communicate. Traditional machinery, as well as new smart gadgets, is "joining the Internet," allowing for the development of more integrated services and the optimization of current systems.

2.2   Cyber-Physical Systems

Cyber-physical systems (CPS) are transformational advances for overseeing connected frameworks counting both physical and computational resources. The competitive nature of today's commerce drives more producers to move toward applying high-tech approaches, much obliged to later developments that have come about in way better accessibility and reasonableness of sensors, information collecting frameworks, and computer systems. As a result, the ever-increasing utilization of sensors and organized gear has come about within the nonstop creation of expansive sums of information, named Huge Information. CPS may be advanced created in such an environment for taking care of Enormous Information and utilizing machine interconnectivity to realize the objective of brilliantly, strong, and self-adaptable machines.

Besides, by combining CPS with existing mechanical forms such as fabricating, coordination, and administrations, today's manufacturing plants can be changed into Industry 4.0 offices with awesome financial potential. As seen in Fig. 9, the 5C plan, a recommended 5-level CPS structure, gives a step-by-step system for making and conveying a CPS for mechanical applications. A CPS is made up of two essential useful components: (1) made strides association, which gives real-time information gathering from the physical world and data input from the Internet, and (2) cleverly information administration, analytics, and computing capabilities, which builds the Internet. However, such a need is excessively broad and not explicit enough for universal implementation. The 5C design given here, on the other hand, clearly describes how to build a CPS from the first data gathering through analytics to the end value production through a sequential procedure [13].

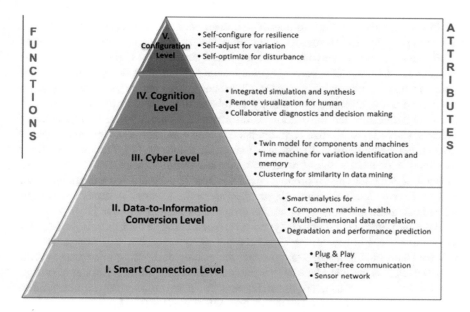

Fig. 9 5C architecture for cyber-physical systems in manufacturing

2.3 Cloud Computing

Following the introduction of personal computers (PCs) into our daily lives in the 1980s, we began to personalize our PCs to meet our software requirements. Despite the fact that they let you to own your own hardware, they have a drawback when it comes to cooperation and sharing. Following that, terminal usage has grown in popularity, since it provides a central repository as well as the ability to run programs locally. Updating hardware components and resolving faults may be a difficult task. As a result, the site of computing is moving again in today's trend. Instead of purchasing hardware, people are now more likely to rent services through the Internet. It may also allow for greater mobility and collaboration. This cloud computing concept also has an impact on the industry or commercial strategies for conducting business through the Internet. Its advantages in terms of manufacturing enterprises are briefly stated.

A cloud is a networked location where computing resources such as computer hardware, storage, databases, networks, operating systems, and even entire software applications are instantly and on-demand accessible. While cloud computing does not introduce much new technology, it does represent an entirely new way of managing information technology that cannot be overlooked. For instance, cloud computing may offer the greatest amount of scalability and cost savings [14].

Industry 4.0 encourages the integration of information and communication technologies (ICT) into production processes to provide personalized goods that meet the high demands of new customers. The Industry 4.0 concept combines operational technology (OT) and information technology (IT) to change the old pyramid model of automation into a network model of linked services (IT). By linking systems and sharing data, this new paradigm enables the formation of ecosystems, allowing for more flexible industrial processes. Cloud computing and big data are important technologies for utilizing the strategy in this setting. The fourth industrial revolution is defined by the integration of information technology into manufacturing processes in order to digitize components and elements, as well as the integration of Operations Technology (OT) and Information Technology (IT). The method allows machines to connect with one another and the use of consumer data in the manufacturing process, allowing for product customization. In addition, the entire production process is completely traceable. It enables interactions from numerous devices and allows machines and humans to connect regardless of their location [15].

Although there are numerous types of research and approaches, it is proposed to assist enterprises in developing cloud computing systems; it is still extremely difficult for traditional industries to construct their cloud computing integrated system; possible reasons include a lack of an integrated cloud computing information system, constantly changing user requirements and business process flows, and vendors lacking the necessary resources and technological capability [16].

Cloud computing promises to provide all of the functionality of present information technology services (as well as allow previously unfeasible new functions) while drastically lowering the upfront computing expenses that prohibit many enterprises from installing many cutting-edge IT services [17].

Cloud computing, in particular, provides the following significant benefits:

1. It lowers the barrier to entry for smaller businesses interested in utilizing compute-intensive business analytics, which was previously reserved for larger businesses.
2. It enables consumers to gain virtually instant access to hardware resources with no upfront financial commitments, resulting in a shorter time to market in a variety of industries.
3. Cloud computing helps lower IT barriers to innovation, as evidenced by the proliferation of intriguing startups ranging from ubiquitous web services such as Facebook and YouTube to more specialized applications such as TripIt (for travel planning) and Mint (for managing personal finances).
4. Cloud computing enables businesses to respond to customer demand by expanding their services—which are becoming increasingly reliant on accurate data. Because computing resources are managed by software, they can be rapidly deployed in response to new requirements.
5. Cloud computing also enables the development of previously unimaginable types of applications and services. Examples include (a) location, environment, and context-aware mobile interactive apps that respond in real time to data provided by human users, nonhuman sensors (such as humidity and stress sensors within a shipping container), or even independent information services (e.g., global weather data); and (b) parallel batch processing, which enables users to leverage

massive amounts of processing power to analyze terabytes of data in relatively short periods of time while programming asynchronously.

One of the most exciting aspects of cloud computing is its potential to assist developing countries in gaining access to information technology without incurring the high upfront costs associated with previous initiatives. Indeed, cloud computing has the potential to do for computing in developing countries what mobile phones did for communications, enabling governments and small businesses to reap the benefits of effective information technology use [17].

2.4 Artificial Intelligence

Another significant issue is artificial intelligence. The journey began with chatbots and has since broadened to encompass a variety of applications. As a result, intelligent manufacturing is a term that refers to the marriage of artificial intelligence (AI) and manufacturing or related product technologies. Artificial intelligence (AI) is the ability of a computer or a computer-controlled robot to perform tasks normally performed by intelligent people.

Additionally, artificial intelligence is critical in assisting the operation of a highly automated system by providing it with increased autonomy in decision-making and defining digital services [18].

"It is common to divide AI into two types: "strong" and "weak" AI (see Fig. 10; [19]). The first would be capable of performing human-like cognitive tasks. Weak or limited AI focuses on narrow tasks and follows predetermined rules. They can accomplish a level of perfection for a one-of-a-kind assignment that would be impossible for a person to do" [20].

As a result, strong AI refers to a system that can generate intelligent behavior that gives the sense of self-awareness and comprehension of one's own reasoning (self-learning). The concept of weak AI refers to an engineering approach to the development of autonomous systems or algorithms that can solve issues.

As a result, the machine imitates intelligence and appears to operate intelligently.

Many manufacturing machines have sensors that can measure their data, as well as data from the environment, changes, and nonconformities. The system used to present the notion of (multi-agent) factor in artificial intelligence is in the form of solving the problem that develops throughout the process. Agents use search, reasoning, planning, learning, symbolic methods, classical and quantitative decision theory models, and symbolic approaches to resolve issues based on the real problem and available technology.

More artificial intelligence aspects will develop as more data becomes available. Things will become more intricate, resulting in a shift from artificial intelligence to machine learning. As the problem grows more complex, transitioning from machine learning to deep learning will become increasingly challenging. It is preferable if you have a lot of data.

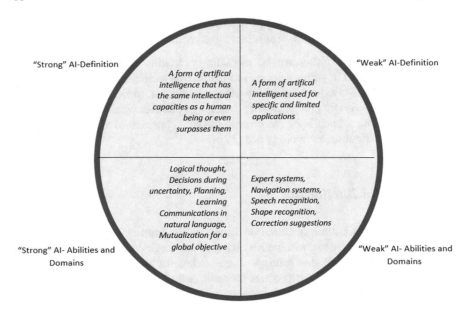

Fig. 10 Artificial intelligence "strong" and "weak"

Deep learning is a machine learning approach that predicts outcomes from a set of data using many layers. Deep learning has aided object detection in pictures, video labeling, and activity recognition and is making significant inroads into other domains of perception, such as audio, voice, and natural language processing.

By 2025, the return for digital projects will be driven by the potential to apply AI to improve decision-making, rethink business models and ecosystems, and redesign the consumer experience [21].

2.5 Service-Oriented Architecture

The Service-Oriented Architecture (SOA) design framework is used to facilitate rapid and cost-effective system development, as well as system flexibility and overall quality. SOA is primarily used in corporate information systems, but it is rapidly gaining traction as a strategy for enterprise information technology systems. It is built on the foundation of web services standards and technology [22]. The SOA is a word used to describe a new generation of IT systems that have unique capabilities in terms of power and flexibility. SOA is built on a modular basis, allowing components of the IT system to be assembled and decomposed to meet the demands of diverse business users [23]. The current business climate is fast changing. Business and product lifecycles are becoming shorter. Globalization, deregulation, entry from different company sectors, and corporate mergers and acquisitions all necessitate business reform or revolution. In addition, the advancement and diffusion of IT (Information Technologies) and the Internet, such as mobile phones, RFID (Radio Frequency IDentification) tags, home information appliances, and so on, create new business opportunities such as online music distribution, electronic cash, automatic acceptance inspection, location-aware item descriptions, and so on. A flexible enterprise information system is required to implement and support the regularly changing business model [22].

To meet the aforementioned business need, information systems must have the following qualities, as illustrated in Fig. 11:

(1) Adaptability to diverse changes: Information systems are now supporting the business models. As a result, information systems must be adjusted in accordance with business models in order to realize the modification or development of business models.
(2) Improving overall system quality: A freshly developed information system frequently contains faults or unsatisfactory functionalities. Defects in information systems have serious consequences for businesses, such as lost opportunities, dead stock, and customer unhappiness, to name a few.

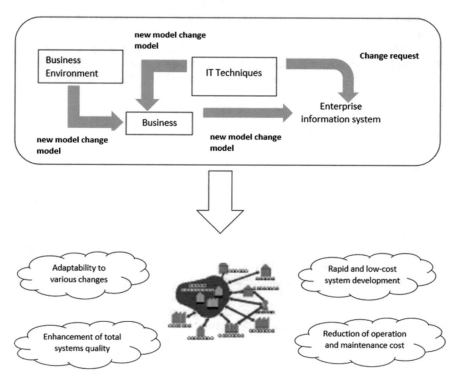

Fig. 11 Change of business environment

(3) Reduced operation and maintenance costs: The majority of lifetime costs are spent on operations and maintenance. Because of the constant changes in the environment, this expense rises. Strenuous efforts were made to achieve the above-mentioned traits. The process-oriented approach first surfaced in the early days. Then, there was the data-driven approach. The object-oriented approach was popularized in the 1990s. System development approaches advance in accordance with not just the computer's and network's capabilities, but also the demands of the environment. These techniques are insufficient from a business standpoint since they are created from a computer standpoint. Based on Internet popularizations, Service-Oriented Architecture (SOA) has gained traction in the twenty-first century as a paradigm that embodies the above-mentioned qualities.

The following advantages are envisaged as a result of implementing SOA.

(1) Services may be simply replaced, allowing the system to be easily updated.
(2) Rapid and low-cost system development is achieved via the use of a variety of services. (reuse, independent system development, standard application)
(3) Using genuine or proven services, whole system quality may be improved and homogenized. It is also simple to isolate a mistake [22].

The potential for SOA and cloud technologies to improve industrial production and control is enormous. SOA enables plant functionalities to be improved and automated, as well as the design, implementation, and aggregation of services. The implemented SOA's primary functional side is represented by a service. The plant communicates with clients and other services through a series of services (Bundzel and Zolotová, Service-oriented architecture and cloud manufacturing, 2016).

2.6 Capability and Competency

Competence is related to capabilities in previous definitions and models. As a result, the terms "competency" and "capacity" are frequently confused [24].

Competency may be defined in a number of different ways. Competence is a widely used term that has diverse meanings for different people. The term "competence," on the other hand, is commonly recognized to relate to the knowledge, talents, attitudes, and behaviors associated with great work performance. Core competences were characterized in the early 1990s as "collective learning within the organization, especially how to coordinate varied production skills and integrate many streams of technology." [25].

Capabilities are defined in a way that takes into account the resources available. Finally, competencies are built on the foundations of resources and capacities [24].

In conclusion:

(1) Resource, activity, and manufacturing strategy are the three key components of capacity modeling.

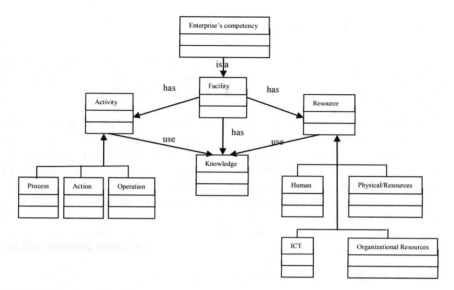

Fig. 12 Enterprise's competency model

(2) Manufacturing strategy has been characterized as "knowledge about processes and resources" in previous studies.
(3) Capabilities are the foundations of an organization's competences.

The goal of the competence modeling framework is to assist model developers in determining what types of competency frameworks are required for good decision-making [26].

As illustrated in Fig. 12, competency is defined as a collection of information and knowledge expressed through all available resources and their associated actions, as well as an understanding of how these resources and activities can be used effectively, efficiently, and economically [27].

While I4.0-related technologies have garnered considerable attention, management techniques tailored to this new ecosystem of connected smart factories have garnered less [28].

Organizations require a comprehensive competency model to successfully manage internal resources/activities and their interrelated activities [29].

When these competence models are paired with organizational goals, they assist to set the performance criteria against which to measure an individual's efficacy and accomplishment in a given job as well as the performance of a team and a company [29].

3   Industry 4.0 Ecosystem

Industrial revolutions are the consequence of a series of inventions that pause and accelerate the march of economic development throughout time. The effects of these innovations evolve throughout time, with the production of subsidiary and supplementary inventions [30].

With the First Industrial Revolution, the industrial technical paradigm that supplanted manual labor was developed. The core of the systemic shifts in industry that occurred throughout this revolution is reduced to the establishment of industrial production. The consequent improvements in logistics included the introduction of steam transport and the production of cast iron items [31].

In the beginning of the 1900s, Henry Ford studied at transition to mass production system from the customize production system, because this approach aims to both finding spare parts and to reduce costs. Thus, these aims were accomplished. The most important output of the second revolution was that to standardize spare parts of the product and to occur the new market in there. In 1974, the third industrial revolution had started with information technologies (IT). With this technology, production entirely transform to automation system. So that, mass customize production system had born.

The Fourth Industrial Revolution is like a Digital revolution where technologies are connected with machine to human or machine to machine. By the way, it includes some differences from other industry revolutions due to using lots of tools or technological knowledge such as Internet of Things (IoT), cloud computing, and cyber-physical system (CPS), in that. This revolution aims to reduce to cost, improve to performance and capacity, operate to conditions faster, and accurate for the increase effectiveness in consideration waste management and saving the energy [32, 33].

Industry 4.0 ecosystem can be described in four main stages like other ecosystems. These stages are birth, expansion, leadership, and self-renewal or death. The birth stage is when actors concentrate on identifying their value proposition (innovation) and how they will interact. When the ecosystem develops to new levels of competition, it enters the second stage, expansion. In the third stage, leadership and ecosystem governance are defined, and leading manufacturers must expand control by guiding major customers' and suppliers' future orientations and investments. Finally, in the last stage, mature ecosystems are endangered by the emergence of new ecosystems and technologies. These dangers have two possible. The ecosystem will either self-renew or perish. As with any other innovation ecosystem, Industry 4.0 innovation ecosystems will face lifecycle stages. This was a critical factor in the German Industry 4.0 initiative's success [34]. In addition to the tools mentioned previously about the industry 4.0 ecosystem, it includes CPS, IoT, cloud computing, and artificial intelligence. Thus, this frame should be designed to function continuously while also being secure against all external threats. Blockchain technology is capable of assisting in precisely this situation. By the way, blockchain is the primary enabler of the Internet of Transactions, which is required for a wide variety of smart manufacturing applications. This functionality enables manufacturers and their suppliers to significantly reduce the time required to complete and register a business agreement. On a blockchain, smart contracts can be created, agreed upon, and added. As a result, they become legally enforceable without the requirement of formal registration [35], January).

This industry revolution impacted numerous sectors of business, including logistics, waste management, pharmacy, quality, and society. The technique is based on a technologically advanced and software-guided system, with the Logistic 4.0 approach significantly increasing material flow efficiency. It is made possible by today's horizontal and vertical applications of information technology, which provide continuous knowledge of a factory's material flow [36]. Among the innovative techniques included in the ReWaste4.0 Industry 4.0 Vision is the conversion of non-hazardous waste treatment facilities into a "Smart Waste Factory" equipped with digital communication and connectivity between high-quality materials and machines, as well as plant performance [37].

CPS applications in drug manufacturing have the potential to revolutionize the control and management systems of biotechnology and pharmaceutical businesses, resulting in a Pharma 4.0 revolution. Among the developments are high-confidence medical equipment and systems, drug research and development, on-demand manufacturing of pharmaceuticals, 3D-printed medicinal goods, Logistic 4.0 for medication distribution management, and green waste management [38].

Quality 4.0 is the application of traditional quality management techniques to modern technology in order to achieve a new level of excellence at the functional and operational levels. Manufacturers who implemented Quality 4.0 technologies achieved exceptional effectiveness and efficiency in quality management, which increased their market share, stimulated innovation, bolstered their ability to deal with value chain challenges, and increased brand awareness [39].

Nowadays, the present information society theory is evolving into a sustainable super-smart society paradigm as a result of the Sustainable Development Goals (SDGs) and elements of sustainability. Society 5.0 promotes more openness and active engagement in social concerns, as well as fair opportunity for all individuals and the integration of modern technology and society. Because of humans, artificially intelligent machines and robotics, advanced analytics and predictive decision support systems, and the use of high-IT technologies in Society 5.0 with Industry 4.0, the integration of cyberspace and physical space in the ecosystem of business and public government is possible, and it creates an environment with the use of high-IT technologies in Society 5.0 with Industry 4.0, as well as the opportunity to develop knowledge workers' competence [40].

In conclusion, Industry 4.0 ecosystem is very large capacity and to effect several fields according to mentioned above. This revelation came to lots of different innovations and perspective. Because of this, people and governments need to prepare for this innovations and perspective, and should be to develop more effective and more beneficiary approaches in short or medium term.

4   Enterprise Architecture in Industry 4.0 Era

Industry 4.0 (I4.0) is defined by an unprecedented level of communication between industrial components. It is made possible by the existing digitalization of communication and information. On the other hand, I4.0 will necessitate an even greater flow of information than the current state of the art in industrial production. Following that, digitization and communication must become more systemic, rapid, organized, collaborative, and cost-effective [41]. Although the final form of the concept was revealed at the specialized Hannover Fair in 2013, the underlying vision of the Fourth Industrial Revolution was revealed two years prior. According to this theory, "smart factories" are managed by cyber-physical systems (CPS) and perform iterative and simple tasks previously performed by humans. Industry 4.0 represents a potential method for enhancing people's quality of life by increasing workforce productivity and eliminating monotonous and physically demanding jobs. Such automation, integration, and increased efficacy associated with more sophisticated logistics have the potential to mitigate the negative impacts of industrial development caused by humans, and they hold great promise for implementing sustainable development [42].

"Industry 4.0" is a term that refers to three interconnected factors: (1) Digitization and integration of all technical–economic relationships, regardless of their complexity, into complex technical–economic networks; (2) Offer of product and service digitization, and (3) The emergence of new business models. Numerous communication technologies were available at the time to connect all of these human activities. The most promising technologies will be the Internet of Things (IoT), the Internet of Services (IoS), and the Internet of People (IoP) (IoP). Due to these communication technologies, communication entities will be able to communicate with one another (in an Industry 4.0 environment) and utilize data from the system's owner throughout the system's life cycle, regardless of business or country borders.

Every entity within the production–market network will have access to pertinent data. It will be extremely beneficial to all entities, particularly manufacturers, because it will be usable even during the design and testing phases. Industrial production digitization may result in the emergence of entirely new digital market models. Users will be able to forecast the shutdown of certain manufacturing entities based on data (available in the cloud) [43]. All entities involved in the production chain will have access to all necessary data, which will benefit the given entities significantly because participants in the production and commercial chains, such as industrial machine manufacturers and software designers, will be able to develop their products based on prior knowledge of the most recent components that have not been fully developed and tested by the producers.

For the purposes of such extensive chains, leading German companies and organizations developed and published the RAMI 4.0 (Reference Architecture Model Industry 4.0) and Industry 4.0 Component models in 2015. RAMI 4.0's three interconnected aspects resulted in the creation of a three-dimensional graphical model [42].

Industry 4.0 discussions bring together previously disparate interests: industries ranging from process to factory automation with vastly different standards, information and communication technologies, and automatic control, the associations Bitkom, VDMA, ZVEI, and VDI, and the standardization organizations IEC and ISO with their national mirror committees in DKE and DIN. It became critical to develop a universal architecture model as a starting point for discussion of its inter-relationships and details in order to gain a clear understanding of the standards, use cases, and other requirements for Industry 4.0 [44]. As a result, the reference architecture model for Industry 4.0 has been created. (RAMI4.0).

BITCOM, VDMA, and ZWEI developed the RAMI 4.0 model.

They chose to create a 3D model because it needed to depict all of the manually integrated technical–economic properties. The SGAM paradigm, which was developed for use in renewable energy networks, appeared to be an excellent fit for Industry 4.0 applications as well. RAMI 4.0 is a minor modification to the SGAM (Smart Grid Architecture Model). Due to the fact that approximately 15 industrial branches are represented in both the SGAM and the RAMI 4.0 models, the RAMI 4.0 model enables alternative perspectives [43].

RAMI 4.0 is an Industry 4.0 Architecture Reference Model created by Platform Industry 4.0 for the purpose of establishing communication structures and a common language within the factory, complete with its own vocabulary, semantics, and syntax.

Such a language enables the integration of IoT and services in the I4.0 environment, connecting them to the rest of the world.

It is a Service-Oriented Architecture (SOA) that combines IT components to facilitate Industry 4.0's key characteristics, such as horizontal and vertical integration within factories, with products on one end and the Cloud on the other.

This architecture, represented by a three-dimensional map with three axes labeled Hierarchy Levels, Product Lifecycle, and Architecture Layers, addresses Industry 4.0 issues in a structured manner, ensuring that all factory participants can communicate with one another and connecting the entire manufacturing process [45].

Figure 13 shows the three-dimensional model of the RAMI 4.0.

The Levels of the Hierarchy Axis is intended to replace the previously proposed Industry 3.0 concept, in which the infrastructure was based on specialized hardware, limiting its operation; the communication model was hierarchical; and the goods were segregated. The objective of I4.0 is to spread the concept of adaptable machines and systems, distributed functions over the network that enable interaction and communication among all participants, and goods that are viewed as architectural elements. The Lifecycle of a Product Axis addresses assets at every stage of their lifecycle, from concept to development and maintenance, to manufacturing, use, and maintenance. Assets are items that a business owns that have a monetary value, such as a device or piece of equipment.

Axis 3 is responsible for developing the CPS proposal, which will be discussed in greater detail below. It includes the RAMI 4.0 Architecture Layers, which are defined in Table 1 and describe the physical entities of the modeled industrial network, such as devices, equipment, and machines, as well as how they are mapped to their virtual representations as Industry 4.0 Components (I4.0C), which details the properties of a

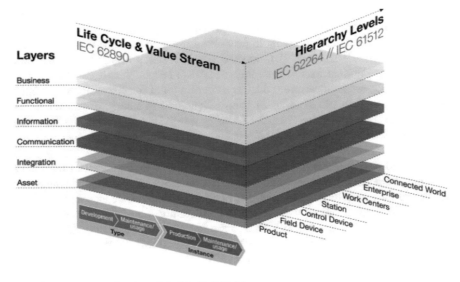

Fig. 13 Three-dimensional model of the RAMI 4.0

CPS. I4.0C are globally and uniquely identified objects with the ability to communicate. They are represented by an asset and an Administration Shell (AS) that provides necessary asset management information. Additionally, it encompasses the technological capabilities of assets, as well as all associated data and information. AS is a standardized network communication protocol that enables the connection of physical objects to Industry 4.0.

The I4.0C can be linked to an asset, machine, or product at the Asset Layer, whereas the AS can be linked to an asset, machine, or product at the Information, Functional, and Business Layers. The AS represents both the physical and digital components of an asset, such as a machine. Additionally, a Superior System Administration Shell (SAS) manages all Administration Shells (i.e., digital twins) in the system by integrating their intercommunication [45].

5 Industry 4.0: Challenges and Opportunities

5.1 Challenges and Issues of Implementing Industry 4.0

While smart manufacturing systems are capable of addressing a wide variety of issues and complexities encountered in established industries, there are still some obstacles to overcome during the installation of smart manufacturing systems [5].

According to several authors, implementing Industry 4.0 will be a challenging task that will take ten years or more to complete. Adopting this novel manufacturing

Table 1 Overview of the architecture layer of the RAMI 4.0

Architecture layers	Description
Asset	The representation of physical objects in the real world. These items include components, hardware, records, and human labor
Integration	Transform the physical world into the virtual one. It reflects visible assets and their digital capabilities, enabling automated event generation and control
Communication	Communication between services and events or data and the Information Layer, as well as between services and control commands and the Integration Layer, is standardized. It is concerned with transmission techniques, network discovery, and their connectivity
Information	The services and data that the technical functionality of the asset may provide, utilize, generate, or modify
Functional	A description of an asset's logical functions, such as technical functionality, in the context of I4.0
Business	Organization of services with the goal of developing business processes and connecting them, as well as supporting business models while adhering to legal and regulatory constraints

technique entails a plethora of considerations and obstacles, including scientific, technological, and economic difficulties, as well as social and political concerns [46]. Challenges of Industry 4.0 are classified as in the Fig. 14.

Other company obstacles and issues include innovation, technology components, digital transformation breakthroughs, and expanding interconnection developments, all of which are significant in any organization. As previously said, Industry 4.0, which entails delivering a new way of manufacturing, is intimately linked to the end-to-end digitization of all physical assets and the integration of all value chain partners into digital ecosystems [46].

The following is a summary of the main problems and issues in the adoption of Industry 4.0 as identified in several studies:

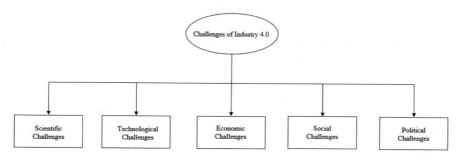

Fig. 14 Different challenges of Industry 4.0 [47]

- The issue of investment is a common one for most new technology-based manufacturing ventures. For an industry to integrate all of the pillars of Industry 4.0, it will require a significant amount of investment [48].
- There is no strategy in place to coordinate efforts across organizational units.
- Talent and competencies are in short supply, such as data scientists.
- A lack of courage in pushing through significant change [49].
- Concerns about the cybersecurity of third-party providers. With Industry 4.0's increased connectivity and adoption of standard communications protocols, the importance of protecting critical industrial systems, manufacturing lines, and system data from cyber security attacks has increased significantly [48]. A smart manufacturing system is one that makes use of an integrated network system to transfer information between manufacturing or machining units and end consumers. This requires network access, which is typically arranged via the Internet. Data and information security are required at numerous stages throughout the system when sharing information over the Internet, including a global unique identity and end-to-end data encryption. As a result, each node in the network should be protected against external threats and data exploitation [5].
- Industry-specific Big Data and Analytics: Ensuring the quality and integrity of data collected from manufacturing systems is challenging. The annotations on data items are extremely varied, making it increasingly difficult to incorporate diverse data repositories with distinct meanings for advanced data analytics [48]
- System Integration: Another challenge associated with implementing a smart manufacturing system is integrating new technology equipment with existing equipment. Compatibility issues between existing and new devices create a slew of complications when it comes to implementing smart manufacturing technology. Certain pieces of machinery that were controlled via specific communication protocols may be obsolete, while new gadgets may use a different protocol. Additionally, machine-to-machine and system-to-system communication necessitate an improved communication infrastructure [5].
- Interoperability: The ability of multiple systems to comprehend and utilize the functionalities of one another independently is referred to as interoperability. This capability enables them to exchange data and information regardless of the hardware or software manufacturer. Interoperability may be difficult to achieve if communication methods and standards are mismatched [5].
- Personnel issues: A shortage of skilled personnel for Industry 4.0, which refers to the demand for employees with cross-disciplinary capabilities in informatics, mathematics, management, data analytics, and engineering in Industry 4.0 organizations. Workers' reluctance to upgrade their knowledge. One of the primary prerequisites for implementing Industry 4.0 is the upgrading of workers' knowledge. On the other hand, employees' aversion to change and upgrading their expertise may prevent the company from initiating or continuing the Industry 4.0 implementation process [50].

5.2   Opportunities of Industry 4.0

The integration of smart products with smart manufacturing, smart logistics, smart networks, and the Internet of Things transforms existing value chains and enables the emergence of new and innovative business models, establishing the smart factory as a critical component of future smart infrastructures. Numerous benefits and profits will accrue as a result of this new infrastructure [46].

According to numerous studies, Industry 4.0 offers numerous benefits to businesses across multiple dimensions. To summarize these benefits, they include the following:

- Advanced planning and control based on current, relevant data
- Quick response to changes in demand, inventory levels, and errors
- Resource efficiency and sustainable manufacturing (materials, energy, people)
- Higher-quality, more adaptable manufacturing
- A rise in productivity
- Reaction to market developments on the fly
- Product personalization
- Increased consumer satisfaction
- Gaining a competitive edge through the successful adoption of a digital business model and the development of technologies
- Cost-cutting and waste reduction
- Workplaces that are safer
- Revenue increase
- The image of a forward-thinking business [51]
- Reduction of waste and overproduction
- Energy consumption can be reduced since energy-intensive jobs can be completed when there is overproduction. For the entire system, energy recovery is used.
- Waste reduction, particularly during the product development phase
- Transportation and travel effort reduction
- Natural resource conservation
- Contribute to current manufacturing plants' environmental aspects [52]
- Products will become more modular and configurable, allowing for mass customization to meet unique consumer needs [53]
- The use of data levering to optimize processes has resulted in a significant boost in all operational efficiencies
- When Industry 4.0 concepts and procedures are used, logistics and data are generated and collected automatically, allowing for quicker responses [46].

Industry 4.0 is considered as one of the primary drivers of revenue growth, despite the fact that its adoption would necessitate large investments on the part of firms. Over the following ten years, the growth it generates will result in a 6% increase in employment. Once Industry 4.0 is fully adopted, productivity in the automobile industry is predicted to grow by 10–20%. For the next five years, operational efficiencies will grow by an average of 3.3% per year, resulting in a 2.6 percent annual cost reduction. In terms of Industry 4.0, revenue will grow quicker and more than the costs of automating or digitizing the production process [46].

All of these factors contribute to higher production and revenue, which can help to boost economic growth [54]. The plant will be able to dynamically adjust its organization and performance levels to suit fluctuating demand thanks to the introduction of I4.0 technology. The production systems will be able to update down/upstream processes (work stations) on their status as a result of the full value chain being networked [55].

5.3 Strengths, Weaknesses, Opportunities, and Threats of Industry 4.0

Industrial advancements make it easier to maximize resource efficiency. Evidently, every business strives to provide the greatest amount of product with the smallest amount of resources. The improved governance system of the fourth industrial revolution allows for exact setting change, which has the consequence of reducing resource usage. Production facilities are not the only ones who benefit from the changes. The human workforce, whose work-life balance can be improved, is likewise affected by flexibility. The clever work structure and design allow professionals to create plans that fit their preferences. The rise of unprecedented job prospects represents the final stage in the human realm. The IT and engineering fields, in particular, will see an increase in demand. Figure 15 highlights all of Industry 4.0's strengths, weaknesses, opportunities, and threats [4].

6 Conclusion

Industry 4.0 is unlike any previous era, most notably the computer era. Industry 4.0 is defined as the convergence of new technologies that blur the distinction between the physical and digital worlds of manufacturing. The rise of cyber-physical systems (CPSs), the Internet of Things (IoT), and Cloud Computing has fueled manufacturing technological advancements that have sparked a new industrial revolution. Additionally, this trend toward industrial digitization is accelerating at a rate, scale, and systemic impact unprecedented in manufacturing history. Industry 4.0's fundamental premise is that new technologies enable organizations to connect people, objects, machines, and systems throughout the value chain in order to create intelligent networks. As the virtual and physical worlds of production collide, hundreds of billions of machines, systems, and sensors will interact, share information, and control one another autonomously. This not only improves production efficiency, but also provides businesses with greater flexibility when it comes to adjusting output to meet customer and market expectations. The Digital Enterprise is the future of

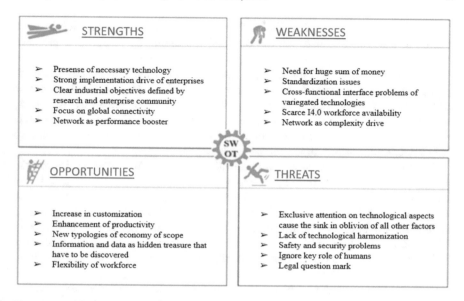

Fig. 15 The SWOT analysis of the industrial 4.0 revolution [4]

manufacturing, and proponents of Industry 4.0 argue that it is the only way for manufacturers to remain relevant and competitive in an increasingly complicated and demanding global market.

Cognitive engineering is widely used in the analysis, design, decision-making, and socio-technical systems. Cognitive physical systems are used in a variety of applications, including human–robot interactions, transportation management, industrial automation, health care, and agriculture.

The primary characteristics of cognitive cyber-physical systems are ubiquitous computing, massive networking, restructuring, a high degree of automation, and interactions with and without supervision [56].

For collaborative work, mutual trust between humans and robots is modeled and computed in real time as an expression of human and machine cognition.

When it comes to HRC in assembly, the human is not merely a beneficiary of the partnership. The human is not merely a conduit for the robot's physical engagement during assembly. Rather than that, it is believed that a human in the HRC during assembly is an active collaborating agent whose cognitive states, as well as physical interactions, can have an effect on the overall system's performance. As a result, it appears as though human cognition must be reflected in the CPSS framework, which can accurately describe the human's involvement in the CPSS [57].

Web 3.0 refers to the next generation of the Internet, in which websites and applications will be able to handle data in a clever, human-like manner through the use of technologies such as machine learning (ML), big data, and decentralized ledger technology (DLT).

According to some futurists, Web 3.0 is also known as the Spatial Web because it aims to blur the line between the real and the digital by reinventing graphics technology and bringing three-dimensional (3D) virtual worlds into sharp relief.

Amazon, Apple, and Google are just a few of the companies that are developing or repurposing existing products as Internet 3.0 applications. Siri and Wolfram Alpha are two Web 3.0 applications. Individuals will not only retain ownership of their data, but will also be compensated for their online time under 3.0. The goal of Web 3.0 is to significantly improve the speed, simplicity, and efficiency of Internet searches, so that even complex search terms can be processed quickly.

The fifth-generation mobile communication network (5G) is an exciting technology that promises not only faster speeds than 4G, but also a service revolution that will impact a wide variety of industries, including health, manufacturing, and energy. Wi-Fi 6 is a new wireless local area network (WLAN) technology that is suitable for use in offices, homes, and densely populated areas. Due to the fact that 5G is a new technology, it has various limitations in terms of coverage and capacity. To circumvent these constraints, one possibility is to utilize the unlicensed spectrum provided by Wi-Fi technology. As a result, a complement solution for 5G/Wi-Fi 6 coexistence is proposed, in which both technologies work in tandem to provide a higher quality of service to the end user in terms of increased speed, lower latency, and increased capacity. Due to the growing demand for higher data rates and shorter buffering times, 5G cellular communication technology will continue to evolve in order to provide end users with services that maximize performance and deploy handsets much faster than previous cellular standards, as existing devices will be incompatible with the new 5G services [58].

5G has the potential to enable significant advancements in automation and artificial intelligence by serving mission-critical applications.

As robots become more intelligent and adapt in real time, they will require massive amounts of processing data that a single machine cannot handle. The solution is 5G-enabled cloud robotics, which centralizes system intelligence and connects it to ground-based robots. The network enables reliable Internet access, remote computing, storage, and data resources, as well as data-driven intelligence, robust cybersecurity, and support for a large fleet of robots. For example, the cloud will serve as a central "brain" in future hospitals, coordinating collaborative robots that advise patients and distribute medications in fleets.

Industrial end-devices such as autonomous guided vehicles (AGVs) are used on 5G networks. Indeed, the mobility management, coverage, and service quality assurance capabilities of 5G networks enable the reliable connectivity necessary for a variety of automated guided vehicles. AGVs include tractors, pallet movers, and forklifts. For example, 5G can assist forklifts in moving more efficiently and automatically across the manufacturing floor.

Connectivity for robots, reduced reliance on fiber and cable tethering, live remote monitoring of robotic video feeds, and low-latency remote control applications are just a few of the ways 5G helps connect production line robotics.

This chapter discusses the revolutions, technologies, and ecosystem of Industry 4.0, as well as the associated challenges and opportunities. Additionally, we discussed the critical role of modeling in Industry 4.0. RAMI was introduced as a Reference Architecture Model for Industry 4.0. The future technologies that will be available are briefly discussed.

With the advancement of technological advancements, inventions, and innovative ideas, the primary focus has shifted to advancing Industrial Revolution 4.0, which is becoming increasingly sophisticated and diverse on a daily basis.

References

1. Kumar A, Nayyar A (2020) si 3-industry: a sustainable, intelligent, innovative, internet-of-things industry. In: A roadmap to industry 4.0: smart production, sharp business and sustainable development. Springer, Cham
2. Hajoary PK, Akhilesh KB (2021) Conceptual framework to assess the maturity and readiness towards Industry 4.0. In: Industry 4.0 and advanced manufacturing. Springer, Singapore
3. Balog M, Trojanová M, Balog P (2019) Experimental analysis of properties and possibilities of application RFID tags in real conditions. In: Industry 4.0: trends in management of intelligent manufacturing systems. Springer, Cham
4. Kumar SA, Bawge G, Kumar BV (2021) An overview of industrial revolution and technology of industrial 4.0. Int J Res Eng Sci (IJRES), pp 64–71
5. Phuyal S, Bista D, Bista R (2020) Challenges, opportunities and future directions of smart manufacturing: a state of art review. Sustain Futures 2:100023
6. Sharma A, Singh BJ (2020) Evolution of industrial revolutions: a review. Int J Innovative Technol Exploring Eng (IJITEE)
7. Pessôa MV, Becker JM (2020) Smart design engineering: a literature review of the impact of the 4th industrial revolution on product design and development. Res Eng Design 175–195
8. Chakrabarti A, Arora M (2021) Industry 4.0 and advanced manufacturing: proceedings of I-4AM 2019. Springer Nature, Berlin
9. Columbus L (2018) Forbes. https://www.forbes.com/sites/louiscolumbus/2018/12/13/2018-roundup-of-internet-of-things-forecasts-and-market-estimates/?sh=3def74cf7d83adresinde nalındı
10. Stoyanova M, Yannis N, Panagiotakis S, Pallis E, Markakis E (2020) A survey on the internet of things (IoT) forensics: challenges, approaches, and open issues. IEEE Commun Surv Tutorials 22(2)
11. HornNord J, Koohang A, Paliszkiewicz J (2019) The internet of things: review and theoretical framework. Expert Syst Appl 97–108
12. Weber R (2012) Evaluating and developing theories in the information systems discipline. J Assoc Inf Syst 13(1):2
13. Lee JB (2015) A cyber-physical systems architecture for industry 4.0-based manufacturing systems. Manufac let 18–23
14. Manvi S (2021) Cloud computing: Concepts and technologies. CRC Press
15. Rana AK (2021) Industry 4.0 manufacturing based on IoT, cloud computing, and big data: manufacturing purpose scenario. In: Advances in communication and computational technology. Springer, Singapore
16. Xia TZ (2021) Using cloud computing integrated architecture to improve delivery committed rate in smart manufacturing. Enterp Inf Syst 15(9):1260–1279
17. Marston SL (2011) Cloud computing-the business perspective. Decis Support Syst 51(1):176–189
18. Pakkala DS (2019) Digital service: technological agency in service systems. In: Proceedings of the 52nd Hawaii international conference on system sciences, vol. 6, pp 1886–1895

19. Paschek D, Luminosu C, Draghici A (2017). Automated business process management—in times of digital transformation using machine learning or artificial intelligence. Timisoara, Romania
20. Péletier B, Nominacher M (2017) Artificial intelligence policies. In: The digital factory for knowledge. ISTE Ltd, London, UK and John Wiley & Sons, New York, USA
21. Echeberria L (2020) A digital framework for industry 4.0. Springer International Publishing. Springer Nature Switzerland AG
22. Komoda N (2006) Service oriented architecture (SOA) in industrial systems. In: 2006 4th IEEE international conference on industrial informatics. IEEE, pp 1–5
23. Ordanini A, Pasini P (2008) Service co-production and value co-creation: the case for a service-oriented architecture (SOA). Eur Manag J 26(5):289–297
24. Vatankhah B, Hashemipour M, Guerra-Zubiaga D (2015) A framework for modelling enterprise competencies: from theory to practice in enterprise architecture. Int J Comput Integr Manuf 28(8):791–810
25. Boyatzis R (1982) The competent manager: a model for effective performance. Wiley Interscience, New York
26. Rosemann M, van der Aalsta W (2007) A configurable reference modelling language. Inf Syst 1–23
27. Barenji R, Hashemipour M, Guerra-Zubiaga D (2013a) Developing a competency-assisted collaborative promotion modeling framework in higher education. Procedia Soc Behav Sci 107:112–119
28. Shet S, Pereira V (2021) Proposed managerial competencies for industry 4.0–implications for social sustainability. Technol Forecast Soc Change 173:121080
29. Barenji R, Hashemipour M, Guerra-Zubiaga D (2013b) Toward a modeling framework for organizational competency. In: Doctoral conference on computing, electrical and industrial systems. Springer, Berlin, pp 142–151
30. Testik M, Sarikulak O (2021) Change points of real GDP per capita time series corresponding to the periods of industrial revolutions. Technol Forecast Soc Chang 170:120911
31. Popkova E, Ragulina Y, Bogoviz A (2019) Fundamental differences of transition to industry 4.0 from previous industrial revolutions. Industry 4.0: industrial revolution of the 21st Century. içinde Springer, Berlin, pp 21–29
32. Alaloul W, Liew M, Zawawi N (2020) Industrial revolution 4.0 in the construction industry: challenges and opportunities for stakeholders. Ain Shams Eng J 11(1):225–230
33. Sony M, Naik S (2020) Industry 4.0 integration with socio-technical systems theory: a systematic review and proposed theoretical model. Technol Soc 61:101248
34. Benitez G, Ayala N, Frank A (2020) Industry 4.0 innovation ecosystems: an evolutionary perspective on value cocreation. Int J Prod Econ 228:107735
35. Mohamed N, Al-Jaroodi J (2019) Applying blockchain in industry 4.0 application. In: 2019 IEEE 9th annual computing and communication workshop and conference (CCWC). IEEE, pp 0852–0858
36. Di Nardo M, Clericuzio M, Murino T, Sepe (2020) An economic order quantity stochastic dynamic optimization model in a logistic 4.0 environment. Sustainability 12(10):4075
37. Sarc R, Curtis A, Kandlbauer L, Khodier K, Lorber K, Pomberger R (2019) Digitalisation and intelligent robotics in value chain of circular economy oriented waste management–a review. Waste Manage 95:476–492
38. Hariry R, Barenji R, Paradkar A (2020) From industry 4.0 to pharma 4.0. In: Handbook of smart materials, technologies, and devices: applications of industry 4.0. içinde Springer, Berlin, pp 1–22
39. Sader S, Husti I, Daroczi M (2021) A review of quality 4.0: definitions, features, technologies, applications, and challenges. Total Qual Manage Bus Excellence, 1–19
40. Sołtysik-Piorunkiewicz A, Zdonek I (2021) How society 5.0 and Industry 4.0 ideas shape the open data performance expectancy. Sustainability 13(2):917
41. Bradac Z, Zezulka F, Marcon P (2018) Technical and theoretical basis of industry 4.0 Implementation. Fundam Sci Appl 13

42. Marcon P, Zezulka F, Vesely I, Szabo Z, Roubal Z, Sajdl O, Dohnal P (2017) Communication technology for industry 4.0. In: 2017 progress in electromagnetics research symposium. Spring (PIERS), St Petersburg, Russia, pp 1694–1697
43. Zezulka F, Marcon P, Vesely I, Sajdl O (2016) Industry 4.0–an introduction in the phenomenon. IFAC-PapersOnLine 8–12
44. Adolphs P, Bedenbender H, Dirzus D, Ehlich M, Epple U, Hankel M (2015) Reference architecture model industrie 4.0 (rami4. 0). ZVEI and VDI
45. Pivoto D, de Almeida L, da Rosa Righi R, Rodrigues J, Lugli A, Alberti A (2021) Cyber-physical systems architectures for industrial internet of things applications in Industry 4.0: A literature review. J Manufac Syst 176–192
46. Mohamed M (2018) Challenges and benefits of Industry 4.0: an overview. Int J Supply Oper Manage 256–265
47. Zhou K, Liu T, Zhou L (2015) Industry 4.0: towards future industrial opportunities and challenges. In: 2015 12th International conference on fuzzy systems and knowledge discovery (FSKD). IEEE, pp 2147–2152
48. Vaidya S, Ambad P, Bhosle S (2018) Industry 4.0–a glimpse. Procedia Manufact 233–238
49. Dennis K, Nicolina P, Yves-Simon G (2017) Textile learning factory 4.0—Preparing Germany's textile industry for the digital future. In: 7th conference on learning factories, CLF 2017 procedia manufacturing, pp 214–221
50. Bajic B, Rikalovic A, Suzic N, Piuri V (2020) Industry 4.0 implementation challenges and opportunities: a managerial perspective. IEEE Syst J
51. Uglovskaya E (2017) The new industrial era: Industry 4.0 & bobst company case study
52. Waibel MW, Oosthuizen GA, Du Toit DW (2018) Investigating current smart production innovations in the machine building industry on sustainability aspects. Procedia Manufac 774–781
53. Pereira AC, Romero F (2017) A review of the meanings and the implications of the Industry 4.0 concept. Procedia Manufac 1206–1214
54. Du Plessis CJ (2017) A framework for implementing Industrie 4.0 in learning factories. Doctoral dissertation. Stellenbosch University
55. Wichmann RL, Eisenbart B, Gericke K (2019) The direction of industry: a literature review on Industry 4.0. In: Proceedings of the design society: international conference on engineering design. Cambridge University Press, pp 2129–2138
56. John A, Mohan S, Vianny D (2021) Cognitive cyber-physical system applications. In: Cognitive engineering for next generation computing: a practical analytical approach. Içinde, pp 167–187
57. Rahman S (2019) Cognitive cyber-physical system (C-CPS) for human-robot collaborative manufacturing. IEEE, pp 125–160
58. Zreikat A (2020) Performance evaluation of 5G/WiFi-6 coexistence. Int J Circuits, Syst Signal Process 904–913
59. Lojka T, Bundzel M, Zolotová I (2016) Service-oriented architecture and cloud manufacturing. Acta Polytech Hung 13(6):25–44

Cyber-Physical Systems—Manufacturing Applications

Ayşegül Kocabay and Hatef Javadi

1 Introduction

Cyber-physical systems (CPS) are a new class of systems that incorporate both computing and physical capabilities and enable people to interact in novel ways. The ability of an organism to interact with its physical environment and hone its skills via computation, communication and control is critical for future technological advancements. Developing next-generation aircraft and spacecraft, hybrid gas-electric vehicles, fully autonomous urban driving, and prosthetic limbs that use brain impulses to manipulate physical objects is one of the opportunities and research challenges. Time and frequency domain approaches, state-space analysis, system identification, filtering, prediction, optimization, robust control, and probabilistic control are advanced systems science and technology methods and tools developed by system and control experts over many years [1]. It is only a portion of it. Simultaneously, computer scientists are developing new programming languages, real-time computing technologies, visualisation techniques, compiler designs, embedded system architecture and system software, and novel approaches to ensuring the reliability, cybersecurity, and resiliency of computer systems. We have made tremendous strides. Additionally, computer science researchers have developed a plethora of robust modelling formats and validation tools. The study of cyber-physical systems is a branch of computer and engineering science (network, control, software, human interaction, learning theory, electricity, machinery, chemistry, biomedicine, materials science, and other engineering). Numerous engineering systems have been built in the industry by decoupling control system design from details of hardware/software implementation. After developing and validating the control system through extensive simulation, ad-hoc tuning strategies were used to deal with modelling uncertainty and random disturbances. However, integrating numerous subsystems while

A. Kocabay (✉) · H. Javadi
Department of Industrial Engineering, Hacettepe University, Beytepe Campus 06800, Ankara, Turkey
e-mail: aysegullkocabay@gmail.com

© The Author(s), under exclusive license to Springer Nature Singapore Pte Ltd. 2023
A. Azizi and R. V. Barenji (eds.), *Industry 4.0*, Emerging Trends in Mechatronics,
https://doi.org/10.1007/978-981-19-2012-7_2

maintaining the system's functionality and functionality can be time-consuming and expensive. For example, in the automotive industry, vehicle control systems comprised multiple suppliers' system components, each with its own software and hardware. OEMs that supply supply chain components face a significant cost reduction challenge in developing components that can be integrated into a variety of vehicles. Component complexity and the use of cutting-edge technology in sensors and actuators, wireless connections, and multi-core computers make developing next-generation vehicle management systems increasingly difficult. Both providers and integrators require new system sciences to enable the integration of individually designed system components in a reliable and cost-effective manner. Theory and tools, in particular, are required to develop the following cost-effective methods: (1) create, analyse, and validate components at a variety of abstraction levels, including system and software architecture, while adhering to constraints imposed by higher levels; (2) analyse and comprehend the vehicle's control system's interaction with other subsystems (engine, transmission, steering, wheel, brake, suspension); and (3) ensure operational safety, stability, and performance. The addition of new features and the cost of vehicle control are becoming increasingly important differentiators in the automotive industry's profitability [2]. CPS (Computer Physical Systems) is a term that refers to a system in which computers and physical processes are combined. Embedded computers and networks monitor and control physical processes. Computers and networks frequently incorporate feedback loops in which physical processes perform calculations and vice versa. In the physical world, time passes inexorably, and simultaneity is unique. None of these characteristics are present in contemporary computing and network abstractions. CPS applications have the potential to outlive the twentieth-century information technology revolution. Among them are reliable medical equipment and systems, life support, traffic control and safety, advanced automotive systems, process control, energy savings, environmental control, avionics, instrumentation, critical infrastructure control (electricity, water resources, communication systems), decentralised robotics (remote presence, telemedicine), defence systems, manufacturing, and intelligent structures. It is not difficult to envision novel features such as distributed micro power coupled to the power grid, where timing accuracy and safety are critical. Transportation systems can greatly benefit from increased intelligence embedded in vehicles, which can improve safety and efficiency. Connected self-driving cars have the potential to significantly improve the military's effectiveness and provide far more effective disaster recovery technologies. Connected building management systems (such as HVAC and lighting) have the potential to significantly improve energy efficiency and demand variability, while reducing reliance on fossil fuels and greenhouse gas emissions. In terms of available bandwidth, cognitive radio can significantly benefit from distributed consensus and distributed control technology. At the right time, financial networks can undergo dramatic changes. The characteristics of distributed real-time control systems may be inherited by large-scale service systems that track goods and services using RFID and other technologies. Distributed real-time games equipped with sensors and actuators have the potential to fundamentally alter the (extremely passive) nature of online social interactions. Each of these applications provides substantial economic benefits. Current computing and networking technologies, on the other hand, may inadvertently obstruct progress towards these applications. For example, today's "best effort" network solutions lack time semantics and sufficient parallel processing models for data processing, making it difficult to achieve predictable and reliable real-time performance. Object-oriented design and service-oriented architecture are two examples of software component technologies that are based on software-friendly abstractions rather than physical systems. Numerous applications are impossible to implement without significant changes to the underlying abstraction [3].

2 Requirements for CPS

Historically, embedded systems have been held to a higher standard of stability and predictability than general purpose computers. Consumers do not anticipate their televisions crashing and restarting. They became increasingly reliant on reliable automobiles, and the introduction of computer control significantly improved the vehicle's reliability and efficiency. Expectations for this level of reliability increase as the transition to CPS proceeds. Unless and until CPS is reliable and predictable, it is not used in applications such as traffic management, vehicle safety, and medical care. However, the physical world is not entirely predictable. Because cyber-physical systems do not operate in a controlled environment, they must be resilient to unforeseen events and adapt to subsystem failures. While designing predictable and reliable components simplifies their integration into a predictable and reliable system, engineers face unique challenges. However, no component is completely reliable, and when unforeseen circumstances arise, the physical world can obstruct prediction. How much can the designer rely on predictability and reliability when developing the system if the components are predictable and reliable? How do you prevent vulnerable structures from collapsing due to slight deviations from intended operating conditions? This is not a novel concept in the field of technology. Digital circuit designers have come to rely on circuits that are highly predictable and reliable. Circuit designers have devised methods for achieving unprecedented levels of accuracy and reliability in the history of human ingenuity, primarily through the use of stochastic processes (electron movements). You can design circuits that perform logical operations billions of times per second almost without error. Each of these is built on an extremely haphazard foundation.

Does this predictability and reliability have to be relied upon by system designers? Indeed, this is at the heart of all modern digital technologies. There is considerable debate over whether this dependency actually retards the advancement of circuit technology. Circuits with extremely small feature sizes are susceptible to the unpredictability of the underlying board, and if the system designer does not place an excessive amount of reliance on the digital circuit's predictability and reliability, the feature size will be smaller. You can easily migrate. Large semiconductor foundries have not yet taken the risk of developing circuit manufacturing techniques that reliably produce logic gates that operate as expected in more than 80% of the time. These gates are considered to be completely failed, and the yields are low when gates are generated on a regular basis [4]. On the other hand, system designers create systems that can withstand such failures. The objective is not to improve the final product's reliability, but to increase the yield. A failed gate is one that has a 20% chance of collapsing and thus should be bypassed by a successful system. Gates that have not failed previously will work in the majority of cases. The implication is not whether a robust system should be designed, but how robust it should be installed. Do you require a system with a gate that operates as expected with an 80% probability? Or do you need to reconfigure the system around a gate that has a 20% chance of failing and build a system that assumes the remaining gates operate at nearly 100% probability? To my mind, the benefit of being able to construct a gate that passes the yield test function nearly completely is enormous. This level of robustness is desirable at all abstraction levels in system design.

It does not, however, negate the requirement for robustness at a higher level of abstraction. Despite the components' high reliability and predictability, storage system developers consider checksums and error correction codes. Even near-perfect reliability can fail when a billion components (such as Gigabit RAM) operate a billion times per second. The principle we must adhere to is straightforward. Components at all levels of abstraction should be predictable and reliable to the extent that this is technically possible. If this is not technically feasible, the next level of abstraction must compensate for the component's lack of robustness. Successful design in the modern era is based on this concept. Theoretically, developing predictable and reliable goals is possible. As a result, we end up with a system that is reliant on it. Wireless connections are more unpredictable and difficult to rely on. As a result, we provide an additional level of compensation by utilising robust coding and adaptive protocols. The obvious question is whether it is theoretically possible to achieve predictability and reliability in a software system. If we confine the term "software" to what is expressed in a simple programming language, the software is fundamentally completely predictable and reliable when based on computer architecture and programming languages. With an imperative, non-parallel programming language such as C, designers can be confident that their machines will operate with near-perfect precision [4].

We encounter difficulties as we progress from simple programmes to software systems, particularly cyber-physical systems. Even the simplest C programmes, when used in conjunction with CPS, do not accurately represent system-critical behavioural characteristics, rendering them unpredictable and unreliable. It can be run without errors and with meticulous semantics, but it will still fail to produce the behaviour required by the system. This may result in a missed deadline, for example. Because time is not a part of the C semantics, it makes no difference whether the programme is late or successfully executed. However, it is critical to ascertain that the system is operating properly. Fully predictable and reliable components become unpredictable and unreliable in the critical dimension. This is an illustration of a failure of abstraction [5].

3 Cyber-Physical Systems in the Context of Industry 4.0

CPS are currently used in a variety of applications, including medical equipment, vehicle safety and assistance systems, industrial process control and automation systems, and support systems for coordinating power supply in order to maximise the use of renewable energy [6]. A CPS is composed of a control unit, typically one or more microcontrollers, that regulates and processes the data acquired by the sensors and actuators required for real-world interaction. Additionally, these embedded devices require a communication interface that enables data exchange with other embedded systems and the cloud. For example, data exchange is the most critical component of CPS because it enables data to be linked and analysed centrally. In a nutshell, a CPS is an embedded system capable of transmitting and receiving data over a network. The term "Internet of Things" refers to computerised physical systems (CPS) that are connected to the Internet.

3.1 Industry 4.0

The term "Industry 4.0" was coined at the Hannover Fair with the announcement of the "Industry 4.0" plan. Following the first industrial revolution's "mechanisation" through the invention of the steam engine, the second industrial revolution's "mass production" through the use of electricity, the third industrial revolution's "digitization" through the use of electronics and computer science, and the fourth industrial revolution occurred. Utilize the cyber-physical system (CPS), as well as the Internet of Things (IoT) and associated services. Germany is a CPS industry leader with more than two decades of experience. The integration of cyber technologies that enable Internet-enabled devices enables cost-effective and efficient innovative services, particularly Internet-based diagnostics, maintenance, and operations. Additionally, it assists you in developing new business models, operational concepts, and intelligent controls, while maintaining a strong focus on users and their unique requirements. Industry 4.0's objective is to establish a digital factory with the following characteristics:

A. Smart networking
    Cyber technologies such as wireless and wired communications, intelligent actuators and sensors, and telecommunications are constantly entwining automated systems and devices, internal logistics systems, and operational resources. This enables you to gain direct access to more complex processes and services. This results in the emergence of entirely new ideas and business models that enable optimal resource utilisation and control.
B. Mobility
    Industrial automation is already advancing with the use of mobile devices such as smartphones and tablets. Allows for independent access to automated system

processes and services across time and space. This provides these systems with a new level of diagnostics, maintenance, and operation [7].

C. Flexibility

Industry 4.0 enables unprecedented levels of customization in the design, diagnosis, maintenance, and operation of automated systems. When developing these systems, you can select the best components, modules, and service providers from a large pool of available components, modules, and service providers. Certain diagnostics can be carried out by the user. Access to "big data" facilitates automation in this case. The data can be recalled, intelligently used, and linked for automated diagnostics if desired. To address the issue of a lack of skills, spare parts can be automatically sourced from the most cost-effective manufacturers.

D. Integration of customers

Customers can adapt their articles to meet specific individual requirements thanks to Industry 4.0. Automated systems in the twenty-first century adapt to the needs and abilities of users of all ages. For instance, modern ticket machines can be operated in a variety of ways to accommodate individuals with varying degrees of disability. People are sustainable, healthy, and mobile as a result of automated systems that support them in all circumstances and stages of life [6].

E. New innovative business models

Production in the future will be diverse and adaptable. There are new processes, infrastructure, and services associated with development. Because the products are modular and adaptable, they can be customised to meet your specific requirements. Industry 4.0 creates a slew of new issues that necessitate extensive research. There are numerous concerns about how to evaluate the reliability and safety of these rapidly developing articles, as well as how they can be certified. Another critical task is data protection and security. You must ensure that your knowledge and privacy are safeguarded and unaffected. To accomplish this goal, new concepts and technologies must be developed to ensure the collaboration of multiple groups and units. Additionally, ethical, legal, and social issues must be reinterpreted.

4 Embedded Systems Foundations of Cyber-Physical Systems

Until the late 1980s, information processing was associated with massive mainframes and massive tape drives. Since then, miniaturisation has enabled personal computers to process data (PC). While office applications reigned supreme, some computers were also in charge of the physical world, typically via feedback loops. Later in the decade, Mark Weiser coined the term "ubiquitous computing" [8]. The term refers to Weiser's prediction that computations (and information) will be accessible at any time and from any location. Additionally, Weiser predicted that computers would be incorporated into everyday objects and go unnoticed. He coined the term "invisible

Fig. 1 Relationship between embedded systems and CPS

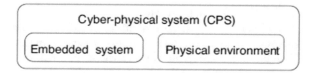

computer" as a result. Due to the anticipated proliferation of computing devices in our daily lives, the terms "pervasive computing" and "ambient intelligence" have been coined to describe similar concepts. Each of the three names denotes a slightly different aspect of future information technology. Pervasive computing is concerned with practical features and the use of currently available technologies, whereas ubiquitous computing is concerned with the long-term goal of making information accessible anytime, anywhere. Future homes and intelligent buildings will place a premium on communication technology that enables ambient intelligence [8]. With the growing use of small devices in conjunction with mobile Internet, some futuristic concepts have already become commonplace. This widespread application is pervasive in the sense that intelligence has an impact on our daily lives. Additionally, miniaturisation enabled computers to integrate information processing into their environment. The term "embedded system" refers to this type of data processing. 1.1 Define (Marwedel [9]) "Embedded systems are information processing systems that are integrated into peripheral products". Embedded systems are used in a variety of applications, including automobiles, trains, aeroplanes, telecommunications, and manufacturing plants. Self-driving cars and trains, for example, are already available or have been announced as embedded system products. As a result, miniaturisation is expected to have the same effect on embedded systems as mobile device availability. Numerous characteristics of embedded systems are shared, including the following: B. Constraints on time, reliability, and efficiency. Connections to physical systems are critical for such systems to function properly. This relationship is emphasised in the following reference [9]. "Software that is embedded in physical processes is referred to as embedded software. The technical difficulty in computer systems is in controlling time and parallel processing". The term "embedded software" is defined in this reference. Can be expanded to include "embedded systems" by simply substituting "system" for "software". However, the term "cyber-physical system" was coined recently. This demonstrates a critical connection to the physical system (CPS for short). Figure 1 illustrates the relationship between the embedded system and its physical environment.

5 Cyber-Physical Systems Security: Limitations, Issues, and Future Trends

Despite their numerous advantages, CPS systems are susceptible to a range of cyber and/or physical security risks, attacks, and challenges. This is due to the fact that it is so diverse, relies on private and sensitive data, and has a large implementation. As a result, intentional or accidental disclosure of these systems is catastrophic and necessitates the implementation of extensive security measures. This can, however, result in an intolerably high network overhead, particularly in terms of latency. Additionally, zero-day vulnerabilities should be mitigated through routine software, application, and operating system updates. Several research papers published recently have concentrated on various security aspects of CPS. Miller and Valasek [10], BouHarb [11], and Sklavos and Zaharakis [12] list and discuss various CPS security objectives. Humayed et al. [13] discuss how to maintain CPS security; Yoo and Shon [14] discuss CPS security challenges and issues; Alguliyev et al. [15] discuss CPS security challenges and issues. Certain security concerns have been examined. Prior research, on the other hand, has not provided a comprehensive overview of CPS security in terms of threats, vulnerabilities, and attacks, depending on the domain of interest (cyber, physical, or hybrid). As a result, this article provides an in-depth examination of contemporary cyber, physical, and hybrid attacks, as well as encrypted and unencrypted security solutions. Additionally, for the first time, CPS forensics is presented as a necessary component of determining the causes of CPS-related crimes and attacks. The CPS system is interconnected with critical infrastructure (smart grids, industry, supply chains, health care, defence, agriculture, and so on) and terrorists. As a result, CPS vulnerabilities can be exploited to launch lethal attacks against such systems. Confidentiality, integrity, and availability are all attackable security features. To ensure widespread acceptance, implementation, and utilisation of these systems, it is critical to safeguard the CPS system against all possible attacks, internal or external, passive or active. The primary objective of this project is to identify the most prevalent CPS security threats, vulnerabilities, and attacks, as well as to ensure that existing security solutions meet safety, accuracy, reliability, efficiency, and security requirements, to describe the advantages and disadvantages of the CPS environment. Additionally, security solutions are evaluated based on their computational complexity. Notably, CPS systems require new security solutions capable of balancing security and system performance.

6   Challenges and Vision

The basic science and technology needed to achieve the CPS goals are crucial to future economic competitiveness. Developing a science and technology foundation for CPS can be successful in many areas of application and lead to extraordinary advances in science and technology. The proliferation of technology brings breakthroughs that bring great social and economic benefits [16]. The key issues with CPS are:

- Electricity generation and delivery without blackouts
- Agriculture with high yields
- In the event of a natural or man-made disaster, a safe and quick evacuation is essential
- Location-independent access to world-class care,

- Near-zero automobile traffic fatalities, low injuries, and considerably decreased traffic congestion and delays,
- Cut the time and costs of testing and integrating complicated CPS systems (such as avionics) by one to two orders of magnitude.
- Energy-aware buildings and towns,
- Physical vital infrastructure that requires regular maintenance, and
- Self-correcting cyber-physical systems for "one-off" applications.

6.1 Cyber-Physical Systems and Design Challenges

When it comes to cyber-physical systems (CPS), there are two important factors to consider. It must first be a physical process. After being detected, sensors can be used to measure various components of this physical process. The data is then processed by the computer system & # 40; "Cyber" & # 41; complete the task. Many CPS now include interfaces for wireless sensors and actuators. With wireless CPS, there are some design issues to address. These questions are explained in Sect. 2.2. Start by investigating the difficulty of bringing together different disciplines to develop such a system. The following describes some of the key KPIs and software tools. The next issue is deciding how to properly represent the sensor readings. Two different algorithms are used to address these issues. The final obstacle is to use network coding for CPS design.

6.2 Challenges for Wireless CPS

Networked CPS research is a difficult area that requires cooperation from different disciplines to advance and overcome the obstacles that arise. Various technical groups need to work together to make this study successful. New system designs often require new computations, metrics, software toolsets, network controls, middleware, and pedagogical foundations. In each of these areas, there are possible solutions to the problem. Calculus faces the daunting task of combining time-based events with event-based systems. The problem with metrics is to develop a mechanism for recording network-oriented metrics related to the system. The software toolset is the next important component. When developing new software, it is difficult to deal with the complexity of cyber and physical interactions. The goal of network control and middleware is to improve the reliability and accuracy of the system, taking into account time sensitivity [17]. Another issue is increasing the number of networked scientists and engineers. New calculations require the exploration of new theories to address the challenges of the future cyber-physical world. To maintain the overall consistency of this broad topic, new ideas are more valuable if they are all based on the same criteria. Manufacturing and testing CPS is a milestone in software development to make your system more efficient. With all these advances, creating enterprise software is a huge step forward for networked CPS control. When training future engineers and natural scientists, the basics of network research need to be emphasised. As a result of growing research and interest in many areas of CPS networking, new engineers and scientists will begin developing newer and better innovations for future wireless CPS.

6.3   Remote Container Monitoring

This section describes programmes that you can use to track and monitor containers on board. As terrorism and smuggling are common battles in the United States, experts are working to build systems that can detect and monitor goods both in ports and at sea. Another consideration in this study is how things move on a daily basis around the world. The ability to perform these tasks remotely is of paramount importance to researchers. Researchers need to use both cellular and satellite cellular networks to complete their work. For the system to function properly, it needs to integrate other applications such as ZigBee, sensors, and RFID (Radio Frequency Identification). All of this works together so users can track the container at any time. Today, there are two techniques for tracking and monitoring containers. Cable transmission is one of them. Use four separate cables and connectors that are carefully placed throughout the container. These sensors use Boolean values to monitor container compression and temperature. Carrier transfer is an alternative option. This technology uses a refrigerated truck modem that sends back frequencies so that the user can check the status of the container. These approaches are the most cost-effective and least time-consuming because they require the effort required to set up. Researchers have discovered a wireless communication solution that can monitor containers more effectively.

6.4   Multilayer CPS Design

This section uses two examples to show the number of layers and heterogeneous attributes used in wireless CPS. These strategies help the system choose the most efficient wireless access network and improve data recovery. As technology advances, so does the way CPS is created, and it needs to keep up. CPS allows users to access the Internet in a variety of ways. For example, you can check the weather using your laptop, mobile phone, or other Internet-enabled device. CPS uses the terminal to get access that can be accessed in different ways. Wireless local area networks (WLAN) and wireless wide area networks (WWAN) are two examples (WWAN). WLAN is an abbreviation for "Wireless Local Area Network", but it can also be connected with a cable [17]. The difficulty is that due to the limited throughput, this is only possible in a limited number of apartments. In most cases, Wi-Fi coverage is limited, but WWAN coverage can reach 90%. To build a CPS cellular network, users must use a heterogeneous multi-layer wireless network that allows both Wi-Fi and

WWAN to be used when Wi-Fi coverage is inadequate. If the WLAN alone cannot support CPS, you need to determine the appropriate way to identify access based on location and available coverage. CPS end devices can only connect to a WLAN if they are within the coverage area of the WLAN. The CPS can connect to the WWAN if it is within the service region of the WLAN. WWAN has much better processing power and speed, but it also costs much more. Algorithms have been developed to enable heterogeneous multi-layer wireless networks. Once the algorithm is in place, you will know what's available and whether Wi-Fi or WWAN meets your needs. The first step in the algorithm is to allocate Wi-Fi bandwidth that does not exceed the CPS end device. The system then assigns a WWAN connection to end devices that are out of WLAN coverage or have inadequate WLAN throughput.

6.5 Energy Efficiency

The following sections provide examples of how wireless CPS supports energy efficiency processes. This is important for designing low cost systems. You can save power by implementing a sleep mechanism or monitoring heat perception and cooling. One of the tasks in this area was planning a wireless sleep using a wireless node. The problem of lowest energy consumption needs to be resolved, taking into account system time. The processor supply voltage and clock frequency are reduced, resulting in reduced energy consumption. The proposed architecture of the reciprocal graph uses a tree topology to achieve these goals. All nodes should be considered when calculating energy savings. Researchers want to achieve this reduction across the network. The idea of putting the radio to sleep when not in use has been proposed as a way to save energy. The problem is that the node does not communicate properly because the node can only communicate when both radios are up. To avoid wasting energy, the radio should only wake up when communication is needed. The recommended treatment is to avoid a set sleep routine. Alternatively, a shared sleep plan with mode assignments was determined to improve the energy efficiency of the entire system. A static power management strategy for your network topology is also required for this method to work properly.

7 Cyber-Physical Systems' Applications

7.1 Cyber-Physical System (CPS) Architecture for Real-Time Water Sustainability Management in Manufacturing Industry

Water, one of the most pervasive but also vital substances on the planet, is necessary for human activity and ecosystems. However, nearly half of the world's population,

3.6 billion people, face water scarcity at least once a year, a figure that is expected to increase to 4.8–5.7 million people by 2050 [18].

It is critical to monitor and evaluate water in this context to ensure sustainable water management. Water footprint assessment (WFA) management is a frequently used technique for water evaluation. This method was also used in the current article's study.

The total amount of freshwater consumed directly or indirectly by human activities over a given time period is referred to as the Water Footprint (WF), which is classified as green, blue, or grey [19]. Green WF represents the volume of non-run-off rainwater used for crop or plant growth, blue WF represents the volume of fresh surface and ground water consumed by human activities, and grey WF represents the volume of "virtual" freshwater required to assimilate the critical contaminant in wastewater in order to meet the requirements of a receiving waterbody.

The WF of an industrial activity comprised two components: supply chain WF (or indirect WF) and operational WF (or direct WF). The WF of the supply chain is not considered in this article's proposed CPS. The article proposes a CPS to ensure water management. The authors began by defining the concept of CPS.

The physical world, interfaces, and cyber system collectively constitute a cyber-physical system (CPS), in which the physical world can interact and work with the cyber system without or with minimal human mediation via the interfaces. The physical world refers to all of the objects that must be observed and controlled in the real world, including human actors, machinery, activities, and the environment. Interfaces, through the use of sensors and actuators, serve as a link between the physical and cyber worlds [20].

As illustrated in Fig. 2, the proposed CPS architecture is composed of four key modules: the physical world, interfaces, the cyber world, and decision-making support. It aims to evaluate and monitor water sustainability for industrial facilities in real time.

The physical world in this application relates to production facilities where fresh water is utilised and wastewater is produced. Interfaces are divided into 2. These are sensors and actuators.

For data gathering, wireless smart sensors for water and wastewater content and volume are required. The smart sensors in the manufacturing plant constitute a wireless sensor network that can collect data from production operations such as the volume of fresh water consumed and the volume and pollutant intensity of wastewater generated.

The actuators in the proposed CPS are alarms and status indicators. If the automatic WF review detects a departure from a pre-set range, the cyber system will sound an alarm. Based on worldwide benchmarks or theoretical factors, the permissible range is set. The system also notifies and guides employees to inspect pertinent processes or instrument in the production facility. As a result, the system offers a real-time dynamic response to modifications in facility condition, assisting in the reduction of economic losses occasioned by excessive water usage and wastewater creation.

The cyber world consists of 4 parts: cloud storage, cloud computing, water footprint monitoring, and decision-making support.

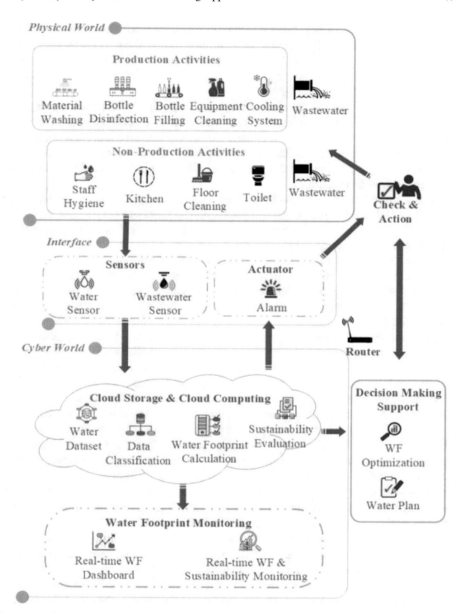

Fig. 2 CPS architecture for water sustainability management [21]

Before further processing, the vast volume of data generated from a manufacturing plant must be automatically preclassified. The data source like the type and position of sensors is used to classify the data. Data from supply-side water sensors, for example, is assigned to the "blue water" group, whereas data from wastewater sensors is assigned to the "grey water" group.

The data is provided to the algorithms in a Cloud-hosted computational engine for estimate the WF for each group after pre-classification. The estimated WF will be compared to a predetermined WF tolerance range based on a local assessment of environmental sustainability. If the result falls outside of the expected range, an alarm will sound, alerting the appropriate personnel. WF calculation and WF evaluation are done in Cloud computing.

The data and computed results are sent to a data archiving and interrogation module, which also control a status dashboard. The dashboard shows the WF and WF sustainability of the entire manufacturing plant, as well as individual processes, in real time. This is called a water footprint monitoring.

As said earlier, water monitoring in the supply chain is not covered in this article. But it is important to monitor water in all sectors to ensure sustainability. Within the scope of the article, the authors predict that this may happen in the future. Thus, water will be monitored from the first step to the last step and transparency and sustainability will be ensured.

A case study in which this CPS can be applied is made in the article. This model has been adapted to a beverage industry. Certain steps have been followed for this. First of all, it is necessary to determine the location of all WF sources in the factory where the production is made. The places of fresh water used and waste water produced should be determined. In addition, these water resources should be classified as green, blue and grey. In the second step, the necessary data for CPS should be determined.

The quantity of freshwater consumed in production operations such as raw material cleaning, bottle disinfection, CIP (clean-in-place) and pasteurisation is necessary for the blue WF computation of a beverage manufacturing plant. Both the proportion of wastewater and the intensity of the "critical pollutant" [19] in the wastewater are required for the grey WF. The definition of a critical contaminant varies depending on the industry and application. Because it is tied to the growth of agricultural or forestry goods, the green WF is foreclosed from the operational WF [19].

The necessary sensors must be placed in the third stage. Appropriate sensors are installed in accordance with the locations of identified water sources. Sensors must be simple to set up and communicate with one another. However, not all sensors could be able to obstruct production. The sensor qualities must be calculated for this. The computation and assessment of water sustainability in the cloud should be supplied for the fourth phase. Some formulations have been devised to help with this. The following are the formulas.

The units of and are measured in cubic metres per hour (m^3/h). The following formula is used to determine the blue water footprint. Where WF is the amount of fresh water (surface water or ground water) abstracted into a procedure or a plant in a unit of time, taking into account water loss and consumption [21].

$$WFblue = \sum WF$$

where CWW denotes the vital pollutant density in the wastewater, WW denotes the wastewater capacity from a process or a plant in a unit of time, CFW denotes the critical contaminant concentration in the abstracted clean water, FW denotes the volume of fresh water abstracted into a procedure or a plant in a unit of time, Cmax denotes the maximum tolerable critical pollutant concentration in the receiving water body, and Cnat denotes [21].

$$WF_{grey} = \frac{C_{ww*}WW - C_{FW*}FW}{C_{\max} - C_{nat}} = \frac{L}{C_{\max} - C_{nat}}$$

Finally, WF is optimised using the CPS outputs. In the short term, unusual water consumption and wastewater generation may be caused by operational and staff errors, machine failures, or pipe leakage. The CPS's alarm capabilities direct necessary personnel to respond appropriately to such situations, resulting in cost savings and reduced environmental impact.

The Cloud-based archived database can be used to create a professional water management plan and a long-term eco-friendly development plan.

This study proposed a cyber-physical-system (CPS) architecture for real-time water sustainability management in the manufacturing sector, in which in-line smart sensing techniques collect data about water and wastewater and cloud storage and cloud computing become accustomed to calculating and evaluating water footprint (WF). A CPS's dynamic features enable industry to monitor and analyse their water footprint sustainability in real time, assisting them in reducing water use and pollution, lowering water costs, and minimising environmental impact [21].

7.2 Cyber-Physical-Based PAT (CPbPAT) Framework for Pharma 4.0

The goal of Industry 4.0 is to achieve smart factories. Indeed, CPS serves this purpose. Machines can communicate with one another and make decisions about problems in smart factories. Numerous businesses are also attempting to transform into smart factories. However, this is not as simple as it sounds, as drug production entails significant risks. On the other hand, the use of continuous production systems and Process Analytical Technology (PAT) in the pharmaceutical industry fostered the adoption of new technologies.

The Industry 4.0 approach enabled by these methods improves the flexibility, cost, quality, and safety of drug production. The Cyber-Physical-based PAT (CPbPAT) framework is proposed in this study. An agent-based approach is used to explain the system.

Quality by Design (QbD), Product Acceptance Testing (PAT), and Real-Time Release Testing (RTRT) programmes have recently gained traction in the pharmaceutical industry to advance product design and manufacturing processes [22]. QbD advocates for a systematic prospective breakdown of product and operation features during the product design process. RTRT and PAT both measure and regulate Critical Process Parameters (CPP) and their relationship to Critical Quality Attributes in a manner similar to "CAM," which is frequently used in typical production systems in other industries (CQA).

Due to the significant hazards and quality requirements associated with pharmaceutical products, they have stringent product and manufacturing operation characteristics that are established using a systematic technique known as QbD [23]. QbD aims to achieve set quality criteria and control product variability by assessing and planning both product and manufacturing process characteristics [24].

PAT is defined as "a system for developing, analysing, and controlling production by performing prompt measurements (i.e., during processing) of critical quality and performance characteristics of raw and in-process materials and processes in order to ensure the quality of the finished product". PAT's primary objective is to enhance manufacturing process control and to apply that information online in order to improve process coordination and, as a result, product quality. According to the definition, RTRT is "a system of release that provides confidence in the product's expected standard, based on information gathered during the manufacturing process, product knowledge, and process understanding and control".

The adoption of QbD, RTRT, PAT, and continuous manufacturing has facilitated advancements in pharmaceutical manufacturing in terms of product quality, cost, time to market, and demand.

Despite the fact that these methods have been developed, current systems are not intelligent. This is for a variety of reasons. To begin, systems are not self-sufficient. They are incapable of making autonomous choices. Additionally, the system is not agile or adaptable enough to deal with a contingency. Finally, the system's interoperability is limited because it is connected to a central computer.

As a result, a cyber-physical-based PAT has been proposed to improve the efficiency of drug production.

The CPbPAT was created with the purpose of collecting, recording, and monitoring data in real time. The CPbPAT analyses acquired data using analytical techniques that enable rapid diagnosis of process, material, and product variability, as well as forecasting of likely sensor and equipment errors. This type of analysis is used to make self-directed decisions and identify opportunities for system improvement. The CPbPAT combines RTRT, QbD, CAM, CPS, and other technologies to provide an independent PAT to drug manufacturing systems. Additionally, the CPbPAT contributes to the development of a more modular, attachable, scalable, flexible, reconfigurable, and adaptable system [25].

The CPbPAT framework's requirement analysis phase examined the system on three levels.

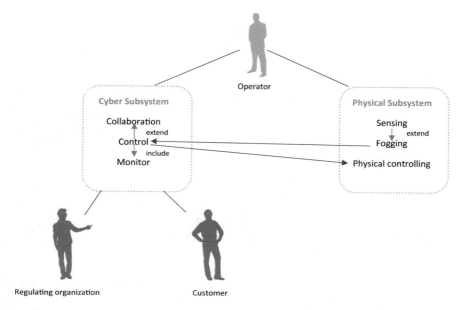

Şekil 1 Use-case diagram of the CPbPAT

1. System level

   At the system level, Fig. 1 illustrates a recommended CPbPAT use-case diagram. The use-case diagram is a UML tool that summarises and visualises both the external interactors (i.e., actors) and the internal use cases of the system. CPbPAT is composed of two subsystems (cyber and physical) and three actor categories (i.e., operator, customer, and regulating organisation). Due to the self-contained nature of CPbPAT, the actors will have no direct control over any of the system's use cases. The characters can only keep track of the characteristics and state of the use cases (Şekil 1).

2. Cell level

   The activity diagram for the cell level is shown in the figure below. The round shape represents the starting node. The cyber interface asks the physical system to work if everything is good. First, the data is checked and sent to the fogging processor. Here, the filtered data is sent to the cyber processor. A decision is made here. If the data is within the predetermined QbD range, the process continues. A device is powered on. Outputs are checked and filtered. And the process ends. If the data is not within the predetermined QbD range, the feeding is stopped and the lot is changed. The process is started again for the new batch. IMA means input material. OMA means output material (Şekil 2).

3. Station level

   The distribution diagram below illustrates the station level. TCP/IP (Transmission Control Protocol/Internet Protocol) is used to connect embedded machines to cloud processors via a fog processor at a single station. The figure depicts the

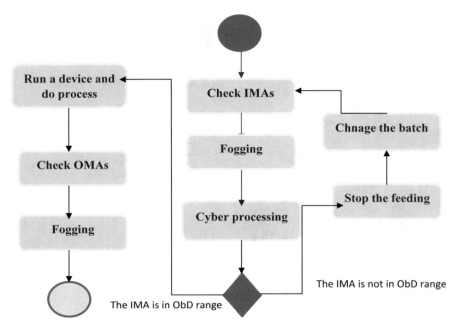

Şekil 2 The activity diagram of the CPbPAT

characteristics of an embedded machine. A cloud processor is a cyber subsystem that uses artificial intelligence to make autonomous decisions. Additionally, it saves and shares data. A fogging processor connects an embedded machine to the cloud processor (Şekil 3).

So far, the passive elements of the system have been represented. An agent-based system is proposed for dynamic models. Agents represent active items. Agents act on what they perceive. Such a system is one that uses common resources, takes place simultaneously, and has several agents communicating with each other.

4. Agent-based framework for CPbPAT (Şekil 4)
   This framework includes four levels; RTRT level, physical system level, fogging level, and cyber-based process level.

Agents within this scope make decisions by communicating with each other. Agents can show 3 different types of features such as decision-making, data interface, and data storage.

It examines the data such as material qualities and quality data from the RTRT level sensors and transfers them to the fogging-level processor. Also, the RTRT level monitors the system. It does this simultaneously with the production process. It provides charts to customers, regulators, and users about the status of the process.

A lot of data is generated in the whole production process. This is where fogging-level agents come into play. These agents first convert the data from the sensors into

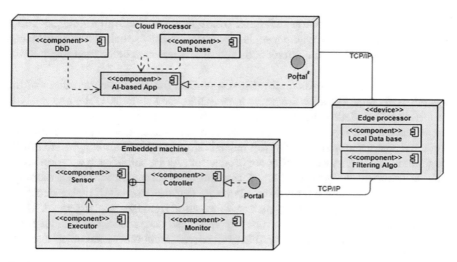

Şekil 3 UML deployment diagram of the CPbPAT [26]

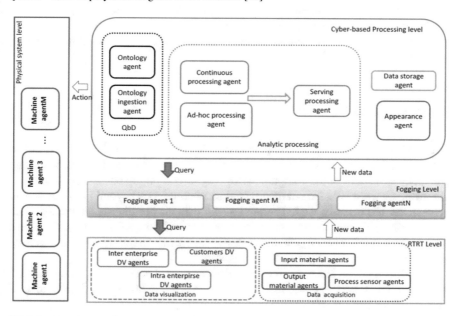

Şekil 4 The CPbPAT framework [27]

the appropriate format and transmit them to the cloud server with appropriate communication protocols. The fogging level is responsible for tasks such as data processing, data collection, and bandwidth reduction. It also filters data before sending it to the cloud.

The cyber-based processing level is responsible for the self-organisation of the system. This level includes 4 agents; appearance, data storage, QbD, and analytic processing. The data that comes out of the filtering comes to this level with the appearance agent. Decisions from this level are also transferred to other levels with the same tool. Data is also saved with the data storage agent. Ontology agent (QbD) contains the necessary information for production. As a result of each decision taken, new information is formed. This information is also transmitted to the ontology ingestion agent. Analytic processing agents are decision makers. It makes decisions with data from the appearance agent and information from the ontology agent. These agents operate in either permanent or temporary (ad-hoc) mode. Continuous mode provides decisions made in predetermined situations. Temporary mode takes decision when unexpected situation occurs.

The physical system level is the nearest to the system's processing equipment. A machine agent is in charge of relaying remarks from the cyber-based processing level to the system's processing equipment.

An example application is also included in the article. For detailed information, you can review the article.

8 Conclusions

Cyber-physical systems have the potential to lay the groundwork for a new era of computing. These systems should be able to achieve new levels of performance and efficiency as a result of the comprehensive control-computing co-design. To accomplish this, we must broaden our understanding of computers beyond information and cyberspace. Historically, we fed computers our data via pre-digested keystrokes and mouse actions. Cyber-physical systems interact in real time with the physical world and consume physical energy in the process. This requires a rethinking of computing as a physical act, a paradigm shift in computing.

Additionally, CPS has been recommended for several years, and research on CPS is just getting started. Numerous impediments to CPS analysis and design arise as a result of the dynamic physical world and the complexity of the cyber world, such as storage constraints, resource constraints, and network bandwidth. This chapter discusses the fundamentals of CPS architecture and examines three critical components: real-time control, security assurance, and integration mechanisms. Additionally, this chapter discusses the current state and progress of cyber-physical systems, as well as their future research directions when applied to manufacturing. The properties of CPS are discussed, as are those of SoS, IoT, Big Data, and Cloud technology. Among the initiatives discussed briefly are Industry 4.0 and the Industrial Internet. In the authors' humble opinion, CPS research and applications will continue in the years ahead, not only because of the unsolved challenges, but also because of the intricate and compelling nature of the problems, which has never failed to fascinate and challenge researchers. This is particularly true when CPS are used in future manufacturing, where self-organising manufacturing, context/situation-aware control, and symbiotic human–robot collaboration can help transform today's manufacturing shop floors into safer and more secure future factories. CPS's unmatched capabilities in networking, communication, and integrated device control contribute to the smartness and intelligence of future manufacturing.

As discussed in the chapter, CPS can be beneficial in a variety of fields, from production to service. It enables machines to make their own decisions by reducing human error. This may enable us to achieve zero error. Artificial intelligence-enabled systems that can evolve continuously will be critical in achieving these goals. CPS recommendations are still in their infancy in transferred applications. Each of them could be improved. CPS, which will provide numerous benefits such as cost savings, optimization, planning, and error reduction, has the potential to fundamentally alter our world. In this section, we attempted to address numerous questions, such as what is CPS, what are its characteristics, and what are the implementation challenges. Additionally, we attempted to create a frame using the two transferred applications. We hope it will be beneficial and serve as a foundation for future CPS applications.

References

1. Krishna PV, Saritha V, Sultana HP (2017) Cyber physical control systems. In: Artificial intelligence: concepts, methodologies, tools, and applications. IGI Global, pp 2575–2599
2. Iordache O (2017) Implementing polytope projects for smart systems. Springer International Publishing
3. Bakakeu J, Schäfer F, Bauer J, Michl M, Franke J (2017) Building cyber-physical systems–a smart building use case. Smart cities: foundations, principles, and applications, pp 605–639
4. Lee EA (2008) Cyber physical systems: design challenges. In: 2008 11th IEEE international symposium on object and component-oriented real-time distributed computing (ISORC). IEEE, pp 363–369
5. Moheyer MG, Ramos MA, Sandoval ER, Pitalúa-Díaz N, Asomoza R, Romero-Paredes G, Matsumoto Y (2021) Notable changes in the performance of a photovoltaic system due to the dirt and cleaning cycles of PV-array. In: 2021 IEEE 48th photovoltaic specialists conference (PVSC). IEEE, pp 0656–0659
6. Jazdi N (2014) Cyber physical systems in the context of industry 4.0. In: 2014 IEEE international conference on automation, quality and testing, robotics. IEEE, pp 1–4
7. Terrissa LS, Meraghni S, Bouzidi Z, Zerhouni N (2016) A new approach of PHM as a service in cloud computing. In: 2016 4th IEEE international colloquium on information science and technology (CiSt). IEEE, pp 610–614
8. Marwedel P (2021) Embedded system design: embedded systems foundations of cyber-physical systems, and the internet of things. Springer Nature, p 433
9. Marwedel P (2021) Evaluation and validation. In: Embedded system design. Springer, Cham, pp 239–293
10. Miller C, Valasek C (2013) Adventures in automotive networks and control units. Def Con 21(260–264):15–31

11. Bou-Harb E (2016) A brief survey of security approaches for cyber-physical systems. In: 2016 8th IFIP international conference on new technologies, mobility and security (NTMS). IEEE, pp 1–5
12. Sklavos N, Zaharakis ID (2016) Cryptography and security in internet of things (iots): models, schemes, and implementations. In: 2016 8th IFIP international conference on new technologies, mobility and security (NTMS). IEEE, pp 1–2
13. Humayed A, Lin J, Li F, Luo B (2017) Cyber-physical systems security—a survey. IEEE Internet Things J 4(6):1802–1831
14. Yoo H, Shon T (2016) Challenges and research directions for heterogeneous cyber–physical system based on IEC 61850: vulnerabilities, security requirements, and security architecture. Futur Gener Comput Syst 61:128–136
15. Alguliyev R, Imamverdiyev Y, Sukhostat L (2018) Cyber-physical systems and their security issues. Comput Ind 100:212–223
16. Rajkumar R, Lee I, Sha L, Stankovic J (2010) Cyber-physical systems: the next computing revolution. In: Design automation conference. IEEE, pp 731–736
17. Patterson C, Vasquez R, Hu F (2013) Cyber-physical systems: design challenges. Cyber-Phys Syst Integr Comput Eng Des 101
18. Water UN (2018) 2018 UN world water development report, nature-based solutions for water
19. Hoekstra AY, Chapagain AK, Mekonnen MM, Aldaya MM (2011) The water footprint assessment manual: setting the global standard. Routledge
20. Ahmed SH, Kim G, Kim D (2013) Cyber physical system: architecture, applications and research challenges. In: 2013 IFIP wireless days (WD). IEEE, pp 1–5
21. Cui X (2021) Cyber-physical system (CPS) architecture for real-time water sustainability management in manufacturing industry. Procedia CIRP 99:543–548
22. Lawrence XY, Kopcha M (2017) The future of pharmaceutical quality and the path to get there. Int J Pharm 528(1–2):354–359
23. US Food and Drug Administration (2018) Guidance for industry: Q8 (2) pharmaceutical development. 2009
24. Lawrence XY (2008) Pharmaceutical quality by design: product and process development, understanding, and control. Pharm Res 25(4):781–791
25. Barenji RV, Akdag Y, Yet B, Oner L (2019) Cyber-physical-based PAT (CPbPAT) framework for Pharma 4.0. International journal of pharmaceutics 567:118445
26. Vatankhah Barenji R (2021) A blockchain technology based trust system for cloud manufacturing. J Intell Manuf 1–15
27. Hariry RE, Barenji RV, Paradkar A (2020) From industry 4.0 to pharma 4.0. Handbook of smart materials, technologies, and devices: applications of industry 4.0, pp 1–22

Internet of Things: Success Stores and Challenges in Manufacturing

Mahmut Onur Karaman, Serap Demir, Şeyda Nur Börü, and Senem Masat

1 Introduction

The Internet of Things (IoT) is a term that refers to the networked interconnection of everyday objects, many of which are equipped with ubiquitous intelligence. The Internet of Things will increase the Internet's ubiquity by integrating every object for interaction via embedded systems, resulting in a highly distributed network of devices communicating with both humans and other devices. Due to rapid advancements in underlying technologies, the Internet of Things is enabling a plethora of novel applications that promise to enhance our quality of life [1]. But, what could be the object in the definition in question? Actually, it can be everything. For example, thanks to the implantable loop recorder placed in the heart, the heart can be considered as a thing, and the information can be transferred to the medical assistant via a server. Or, with the biochip placed on the farm animals, the status of the animals in the farm can be monitored, and the product efficiency can be increased. The farm animals would be thing in this example. Also, automobiles with built-in sensors can be considered as thing. So, various objects can be thing or can transform into a thing.

But, what could be the object in the definition in question? Actually, it can be everything. For example, thanks to the implantable loop recorder placed in the heart, the heart can be considered as a thing, and the information can be transferred to the medical assistant via a server. Or, with the biochip placed on the farm animals, the status of the animals in the farm can be monitored, and the product efficiency can be increased. The farm animals would be thing in this example. Also, automobiles with built-in sensors can be considered as thing. So, various objects can be thing or can transform into a thing.

M. O. Karaman (✉) · S. Demir · Ş. N. Börü · S. Masat
Department of Industrial Engineering, Hacettepe University, Beytepe Campus 06800, Ankara, Turkey
e-mail: onur.karaman@hacettepe.edu.tr

© The Author(s), under exclusive license to Springer Nature Singapore Pte Ltd. 2023
A. Azizi and R. V. Barenji (eds.), *Industry 4.0*, Emerging Trends in Mechatronics,
https://doi.org/10.1007/978-981-19-2012-7_3

Although the Internet of Things first emerged as a term at the end of the 90s, this technology actually dates back to earlier as a prototype. The Coke machine at Carnegie Mellon University, considered the first IoT device, was just a solution of a graduate student, for a problem which he faced in 1982. In 1982, David Nichols, a graduate student at Carnegie Mellon University, got tired of walking to the Coca Cola vending machine only to find that it was out of soda. To his credit, he did say that the Coke machine was "a relatively long way" from his office. Leave it to a computer science student to over-engineer a solution. Nichols and a few friends (Mike Kazar, Ivor Durham, and John Zsarnay) installed micro-switches in each of the vending machine slots. The Coke machine was then connected to the departmental server, and a simple programme was written that could query the switches to determine the day's current state of soda. By typing "finger coke@cmua" from any machine on the Internet (technically ARPNET at the time), the students were able to determine not only if the vending machine had soda available for purchase, but which button would provide the coldest soda based on the time the bottle had been stocked in the machine. (htt2).

2 Structure of Iot

As mentioned above, there are many applications of IoT. With the widespread use of the Internet, it has become possible to connect more objects to each other. Considering that IoT technology is used in a wide variety of fields, it does not sound strange to be able to talk about a standard IoT architecture. In other words, each different IoT application requires its own unique architecture. For example, in the farm example, which is mentioned above, it may be more meaningful to track the data of farm animals with the help of sensors placed on the animal skin. But, when a traffic application or production line is considered, RFID tags may be required to collect data. In addition, if a model for patient follow-up is established, the interface of the application used during the transfer of patient data to the hospital and then sending it back to the medical assistant will be different from the interface of the applications used in smart home systems. But, some research shows some structures which are suitable for most specific applications. These are illustrated below (Table 1).

As a conclusion, each IoT application needs its own unique architecture. However, an IoT application still needs to include some tasks. These are the elements that will enable the application to be worked properly and qualify as IoT. The names and details of these elements are as follows.

2.1 *Identification*

Identification establishes a distinct identity for each object in an IoT application. There are two distinct types of definitions that can be mentioned here. These are

Table 1 Components of various IoT model [2]

Architecture model	Components
European FP7 research project	• Leaves—enable the creations of a maximal set of interoperable IoT systems • Trunk—potentially necessary set of enables or buildings • Roots—interoperable technologies
International Telecommunications Union (ITU) architecture	• Sensing layer • Access layer • Network layer • Middleware layer • Application layer
IoT forum architecture	• Applications • Transportation • Processors
Qian Xiaocong, Zhang Jidong architecture	• Application layer • Transportation layer • Perception layer
Kun Han, Shurong Liu, Dacheng Zhang, and Ying Han architecture	• Near field communication • Network equipment management • High-speed Internet

referred to as naming and addressing. The term "naming" refers to the object's name, whereas "addressing" refers to the object's unique address. These two terms are mutually exclusive, as two or more objects may share the same name but must always have a distinct and distinct address. There are numerous methods for allocating an object's address. Electronic product codes (EPCs) and Internet Protocol version 6 (IPv6) are two examples. Electronic product code is a code that contains information about the manufacturer/manufacturer, the product type, and the product serial number. It is used to identify and track products globally, instantly, and automatically throughout the supply chain. IPv6, on the other hand, is a unique international code that identifies the Internet address of a device.

2.2 Sensing

It is the process of deriving information from objects. The collected data are stored on a storage medium. Numerous sensing devices are used to collect data. RFID tags, smart sensors and wearable sensing devices are just a few examples.

2.3 Communication

The Internet of Things' primary objective is to connect devices. As a result, communication is critical. Radio communication in IoT technology. It incorporates technologies such as radio frequency identification (RFID), near-field communication (NFC), Bluetooth, wireless fidelity (Wi-Fi) and long-term evolution (LTE).

2.4 Computation

Calculation is made in order to remove unnecessary information on the data collected with the help of sensors. There is a lot of hardware and software to perform this process. Hardware platforms are Audrino, Raspberry Pi and Intel Galileo. Software platforms are operating systems such as Tiny OS, Lite OS and android.

2.5 Services

IoT applications provide two basic types of services. The first of these is related to identification. In this type of service, IoT applications receive and transmit the identity of the object. The second type of service is information gathering. It serves to collect information other than address or name from objects. Here are the things to note: the main feature that distinguishes IoT applications from cyber-physical systems is that they do not have a responsibility to make decisions. IoT services do not make a decision from the data they collect in order to apply it to the device.

2.6 Semantics

IoT applications should present meaningful data for users to make decisions. This is actually the most important step of the process. Technologies such as RDF, OWL and EXI are used for this purpose.

As mentioned before, IoT devices should accomplish these tasks for useful applications. On the other hand, each application needs its own architecture. Therefore, a single type of structure cannot be mentioned. However, it can be said that IoT applications should fulfil the above tasks. In addition, IoT devices do not have decision-making purposes and deliver useful data from the object to the decision maker.

Although a single IoT architecture cannot be defined, certain stages of IoT applications are similar in almost all processes. We will call this the layers of the IoT.

3   Layers and Components Iot

Although each IoT application has its own architecture, three key levels are found in all IoT applications. In order to explain IoT technology in the simplest way, it is necessary to explain these layers and its elements. The three layers we will talk about in this context are the sensing layer, the network layer and the application layer (Fig. 1).

3.1   Sensing Layer

The sensing layer is where the IoT application gathers data from the object. Information gathering is a time-consuming process. Because the appropriate information must be gathered for the appropriate purpose, and the objects we use require a variety of applications. A sensor, for example may be required to collect temperature data, whereas RFID tags or barcodes may be required for identification purposes. This is accomplished through the use of a variety of technologies in IoT applications. In a nutshell, these are referred to as sensors.

Sensors in IoT devices perform the function of perceiving information from the object. Diverse sensor technologies have been developed to meet the evolving needs of technology and society. Sensor technologies such as RFID tags, smart things, nanotechnology and miniaturisation and actuators are just a few of them. (Problems

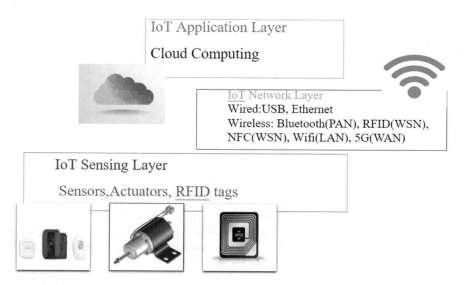

Fig. 1 Layers of IoT

and opportunities with the Internet of Things). This chapter will focus on RFID tags in particular due to their widespread use and promising future.

RFID tags are small devices that come in a variety of shapes and configurations (from stickers to small grains embedded in documents). RFID tags at their most basic level consist of a microchip and a metal coil. The microchip stores data and is capable of performing some simple operations, whilst the metal coil acts as an antenna, receiving data from and transmitting data to readers. RFID tags can optionally include batteries; in this case, they are referred to as active tags. In the absence of that, they are referred to as passive tags. Passive tags are far more prevalent than active tags due to their lower cost. Because passive tags do not contain batteries, they derive their energy from the signal received from readers, resulting in extremely limited storage and computational power. RFID readers are devices that are used to retrieve data from RFID tags. In their simplest form, readers emit a radio wave that powers up and responds to all tags within their coverage range by broadcasting their embedded information (i.e. a set of bits3). After collecting the data, the readers transmit it to a centralised computer (or back end) along with their unique identifier and a timestamp. Back ends are a collection of databases connected to computers that receive, decrypt (as needed) and manage information about RFID tags collected by RFID readers. Back ends maintain all of the data necessary to identify RFID tags. Additionally, they can store additional data about the products/items to which the tags are attached [3].

Actuators serve as alarm and warning systems in IoT applications. Actually, actuators literally mean devices that give a command. However, since the decision mechanism in IoT systems is not given by the model as in cyber-physical systems and IoT systems do not have decision-making responsibilities, actuators in IoT applications are used to warn the decision maker and convert energy into motion. In other words, it is not used to implement a decision made by the IoT system.

There are different types of actuators are used in IOT applications such as DC motors, servo motors and linear actuators.

3.2   Network Layer

At the network layer, the information collected by the sensors is transmitted to the application or storage area by a wired or wireless system. However, when considering IoT applications, although there are wired technologies such as USB and ethernet, the main applications are implemented by using a wireless network. There are different types of wireless network technologies for network layer of IoT applications. These are personal area network (PAN), local area network (LAN), wide area network (WAN) and (WSN).

PAN is an individual organised network in a restricted range area such as within a single building. Bluetooth technology is the popular known example of PAN.

LAN is also a short-range network method. However, it is designed for high data rate transfer from Mbps to Gbps, whilst PAN is used for relatively low data rate. Wi-Fi technology can be given as an example of LAN.

WAN stands for wide area networks. WANs are large and wide networks that are formed by connecting local networks created in different locations. As a result, WANs cover a significantly larger area than LANs. Cities and even countries are large areas. However, in order to connect these LANs, a variety of specialised tools, such as link antennas, gateways, and satellite links, are required.

The WSN system is used to address the requirement for transmitting the data detected by a sensor. As a result, WSN systems are quite prevalent in IoT systems. Due to its low cost and ease of use in terms of energy consumption, WSN technology is a viable option for IoT applications. Wireless sensor network framework agreement, which covers the physical layer, the data link layer, the network layer, the transport layer and the application layer. The physical layer is responsible for carrier frequency, signal modulation and demodulation; the data link layer is responsible for media access and error control; media access protocols enable point-to-point and point-to-multipoint connections in a communication network; error control protocols ensure that the message sent by the source node is complete, correct and reaches the target node, and the network layer protocol is responsible for route discovery and maintenance. Because the majority of nodes and convergence gateways/nodes cannot communicate directly in wireless sensor networks, routing and forwarding via intermediate nodes is required to complete data transfers [4].

RFID is a widely used method of integrating WSN into IoT. RFID (radio frequency identification, Radio Frequency Identification) is a non-contact form of automatic identification. The simplest RFID system consists of three components: the label, the reader, and the antenna; however, practical applications require additional hardware and software support. Its operation is straightforward: the label is placed in the field; an RF signal receiving reader is sent, and the reader reads and decodes the information stored in the chips, Tag, passive tags or passive tags), signal or takes the initiative to send a frequency (Active, tag, active tags or active tags); the reader reads and decodes the information and sends it to the central information system for data processing [4].

RFID technology has a very significant role in IoT technologies due to its cheap cost, less energy consumption and availability for widespread use. Especially, the fact that passive RFID tags do not use any batteries is an important advantage of this technology.

As mentioned above, there are three types of RFID tags as passive RFID tags, active RFID tags and semi-active RFID tags. The main difference between these three types of tags is the battery usage and accordingly the frequency range. Active RFID tags using batteries enable data to be read at a longer range, whilst passive RFID tags allow data to be transmitted over relatively shorter distances.

3.3 Application Layer

At the application layer, the collected data are kept in a storage and presented to the user with an interface that can be understood. In order for the user to access instant data wirelessly, an Internet environment and a storage that can be accessed from anywhere are needed. This need is usually met by cloud computing. Most of today's IoT applications are based on cloud computing. Therefore, in this section, information about cloud computing will be given.

3.3.1 Cloud Computing

Cloud computing is a centralised storage technology that enables the transfer of data and files from users' computers to data centres and back. It is quite simple to gain access to a variety of data and programmes via a centralised cloud system.

Cloud computing is the result of distributed computing, parallel computing and grid computing, which automatically split the large calculating processing programme into several small sub-programmes. Finally, the large amount of data stored in distributed computer products works in conjunction with the processor resources, distributing the relevant calculation to the distributed computer rather than the local computer. It is a business model that provides users with access to IT resources, data, and applications via the network. The purpose of cloud computing is to move independently run, personal calculations from a personal computer or a single server to a "cloud" server with a massive amount of data. This "cloud" server responds to user requests and outputs the results; it is a backbone system based on data computation and processing. Cloud computing can be classified into two categories: service and management (Fig. 2).

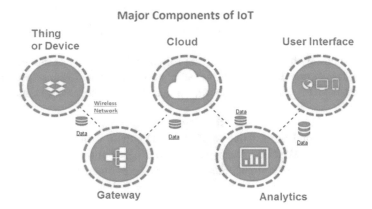

Fig. 2 Major components of IoT. *Source* https://www.globalsign.com/en-sg/blog/internet-things-singapore-futurelandscape

In service, it primarily provides users with cloud-based services; it has a total of three levels. The first layer is software as a service, or SaaS for short; the second layer is platform as a service, or PaaS for short; this layer provides development and deployment of applications as a service to users; and the third layer is infrastructure as a service, or IaaS for short; this layer provides various elements of the underlying computing (such as virtual machines) and storage resources as a service to users. From the user's perspective, these three services are completely independent, as the services they provide are completely distinct and user oriented. However, from a technical standpoint, these three layers are interdependent. For instance, SaaS products and services require not only SaaS layer technology but also the development and deployment platform provided by the PaaS layer or directly deploy on the computing resources provided by the IaaS layer, and the PaaS products and services are likely to be built on top of the IaaS layer services. In management, it is primarily dominated by cloud management; its role is to ensure that the entire cloud computing centre operates securely, reliably and effectively [5].

In an IoT system, devices/things can communicate with one another and with the agent only via the same network protocol. The devices' communication with one another creates a variety of complications. The network protocol that is used to communicate between devices; The network traffic of IoT devices is a concern due to active hardware and power issues related to the amount of data transmitted, data transmission and critical hardware. As a result, various network protocols have been developed by businesses to improve the efficiency with which IoT devices communicate. Increased connectivity within devices does not benefit people in and of itself; however, connecting devices to a network in order to transfer the data they collect from the environment is the most critical step towards successfully utilising the IoT. As a result, cloud computing services have grown in importance day by day.

There are companies today that provide cloud services for a variety of IoT applications. However, when deciding between them, cost-effectiveness, safety and speed should be considered. The market leaders in this space are Amazon Web Services and Microsoft Azure. AWS IoT Core, AWS Greengrass, AWS Kinesis, AWS Lambda, AWS Shield, Azure Stream Analytics, Azure IoT Hub, Azure Service Bus and Azure Security Centre are just a few of the cloud services offered by these two companies for IoT applications.

4 Applications in Iot

Today's technology offers many IoT applications into our lives. Smart cities, wearable IoT devices, IoT-based smart homes and smart devices, healthcare applications of IoT are interests of IoT in daily life (Fig. 3).

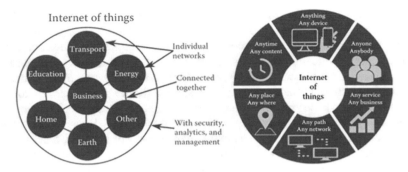

Fig. 3 Applications of IoT [6]

4.1 Smart Cities

As technology advanced, people began to look for ways to improve the quality of life in the cities they live in. In this context, smart city solutions have started to come to the fore. Many areas such as water and waste management, lighting control, energy, transportation, traffic and parking management and building efficiency and security are amongst the research topics of IoT applications for smart cities. Of course, there are real-life examples of smart city applications. Some examples of these are London and Greenwich in the United Kingdom, Barcelona, Murcia and Santander in Spain, Amsterdam in the Netherlands, Aarhus in Denmark, Oulu in Finland, Bordeux in France and a few towns in Korea [7].

Although the widespread use of the smart city concept and the establishment of the infrastructure of all cities with IoT systems may sound futuristic, the concept of smart city is actually quite simple and traditional. The basis of the smart city concept, as stated above, is to bring together new technologies, city infrastructure and government for a joint work in order to improve the quality of life of people. Although current smart city applications are based on smart parking systems, smart roads, smart lighting, smart water and waste management, smart firefighting systems, the opportunities offered by IoT technologies will be used in many more areas in the near future, and the world will become more environmentally friendly and less chaotic [8].

4.1.1 Internet of Things Benefits for Smart Cities

- Significantly increased energy efficiency.
- Enhancement of traffic management.
- Pollution and waste reduction.
- Eliminating crime and enhancing public safety.
- Improving infrastructure management and improving the citizens' quality of life. (htt3).

4.2 Health Care

It is estimated that the economic impact of IoT applications will be greatest in the field of healthcare. Because the health care is an area where more sensors should be used, and more surveillance is required. Realising this, IoT companies can gain a great advantage if they research the field of healthcare systems. Helping doctors in their profession, IoT devices are especially useful in surgery and critical operations. However, IoT systems will also have a very important place in-home care service. IoT devices that are developed for home care services not only help patients feel protected but also bring patient–doctor interaction to a better level. In this way, patient satisfaction also increases. In addition, these IoT devices used in-home care services also contribute to the real-time recovery process and shorten the hospitalisation time of the patients. In the United States alone, more than 200 million people have one or more diseases that require follow-up. This shows how much IoT systems are needed for home care services. Thanks to the new IoT systems, patients can receive the same service at home as they receive in the hospital. Healthcare companies around the world offer an easier life service by using IoT systems that provide home care services to elderly people and disabled people. On the other hand, healthcare institutions also think that IoT solutions are a way to increase the quality of care services and reduce costs [8].

IoT systems can be useful in clinical care and remote monitoring in health care. Thanks to its small size and low cost of biosensors, it can play an important role in the home care service of chronic patients. In this way, the quality of life of patients can be increased and medical intervention can be made more easily in emergencies. As for the elderly people's tracking systems, thanks to IoT applications, elderly citizens can be provided with an independent life as long as possible, and they can be assisted in their daily routines. Considering the ageing of the world population, it can be understood how important elderly care services are. In another example, IoT systems can be used to monitor glucose levels of diabetes patients [7].

4.2.1 IoT Benefits in Health Care

- Decreased wait times and cost savings;
- Improved disease detection and prevention;
- Improved functionality of healthcare devices.
- A decrease in hospital readmissions;
- Improved patient care; and.
- Increased clinic process efficiency (htt3) [8].

4.3   Smart Homes and Smart Buildings

Considering smart home applications, elderly care, control and monitoring of household appliances, energy consumption control and solar energy systems may come to mind. But, the core point of smart home applications is focussed on cost reduction and automation. In this context, heat tracking, light tracking and control, consumption control and energy-saving issues come to the fore. In addition, systems such as fire monitoring and intrusion control systems are also very critical. Intel, IBM and others produce many smart home solutions [7].

Especially, in smart home systems, there are many products offered by IoT. Their general concept was mentioned above. In general, there are air conditioners, speakers, smart thermostats, pet feeders and systems used to improve daily operations in the home. Therefore, smart home systems are currently one of the most popular and widely used areas of IoT, and also, the most promising given their cost-effectiveness and effectiveness [8].

4.3.1   Benefits of IoT in Home Automation:

- Smart energy management and control,
- Centralised management of all home devices,
- Appliance predictive maintenance and remote functionality,
- Enhanced comfort and security,
- Remote control of home appliances and.
- Insights and analytics on smart home management. (htt3).

4.4   Mobility and Transportation

All kinds of vehicles used in a city can become more equipped thanks to IoT applications and sensors and actuators used in this context. In smart traffic systems, roads, rails and vehicles are equipped with RFID tags and other sensors that send information to the traffic control system. In this way, it not only monitors vehicles but also provides better communication of vehicles and makes cities more livable. In addition, thanks to IoT systems, automobile diagnostic services can be used more effectively. Thanks to the sensors on the vehicles, they can collect various information such as pollution, humidity and temperature in the environment. Thus, the concept of an "intelligent vehicle" emerges. Through the collection and evaluation of such data, more environmentally friendly cities can be created. Highways can also be safer. For example, solutions that will make fuel consumption more efficient in public transportation can be produced, and the exhaust emissions of the vehicles can be reduced. In this context, currently used mobile applications such as Google Traffic or Waze* use data provided by the user to monitor traffic conditions. In another traffic application, IoT systems can be used for traffic lights, thus enabling drivers to encounter traffic more smoothly. For example, in an application for cyclists, the infrastructure of traffic lights in the city and traffic data is transmitted to the cyclist's smartphone, and systems can be developed to prevent the cyclist from stopping at every intersection. In another example, a person who orders a good can follow the whole process from the moment the ordered good comes out from the supplier until it reaches him. This is also an example of modern logistics systems [7].

4.4.1   IoT Benefits in Transportation

- Remote vehicle tracking and fleet management.
- Monitoring cargo conditions.
- Improved last mile deliveries.
- Monitoring driver activity.
- Detecting exact vehicle locations and advanced routing capabilities. (htt3)

4.5   Smart Agriculture

As technology develops, modern agriculture needs new requirements to adapt to the competitive environment. In particular, it is very important to produce high yield, quality products and to be ecological. Considering all these, IoT applications contribute to the modernisation of agriculture. Thanks to WSN systems, better services are provided in areas such as irrigation control, animal tracking, fertilisation and pest control. In addition, IoT systems offer various solutions in greenhouse monitoring, viticulture and horticulture [7].

4.5.1   Internet of Things Benefits in Agriculture

- Crop, climate and soil condition monitoring
- Livestock monitoring
- Precision farming
- Watering and fertilisation automation
- Automating detection and eradication of pests
- Greenhouse automation and
- Higher crop quality and better yields. (htt3).

4.6   Wearables

With the help of sensors, the required information is collected from the users. These sensors are wearable, and users wear them. Especially, thanks to those used in the field of health services, heart health can be monitored, or glucose levels of patients can be controlled with a biosensor placed on the skin of diabetes patients. In addition, thanks to smartwatches, a person can access health and fitness information about himself, and with applications such as pedometre, he can reach the number of steps he targets daily and lead a healthier and more comfortable life [8].

4.6.1 Benefits of IoT Wearables

- Remote diagnostics and health monitoring,
- Advanced personal care options for patients,
- Early disease detection and prevention and,
- A data-driven approach to fitness and personal care. (htt3).

4.7 Smart Retail and Supply Chains

Thanks to IoT systems in the retail industry, equipment maintenance, demand alerts and warehouse maintenance can be carried out more efficiently, and all these processes can be monitored more healthily. In addition, other areas where IoT systems can help the retail industry are listed below:

- Advanced supply chain management.
- Better customer service.
- Smarter inventory management.
- Automatic payment.

RFID systems are very useful in this field. Using IoT-connected devices, RFID tags with positioning sensors can help wholesalers and retailers obtain precise information about the location and movement of ordered products from manufacturers to customers, even to details such as when the order was placed in the store when it was delivered. Any problem can be dealt with and resolved quickly in a systematic way. For example, consider products with a short expiration date that requires about high yield, a constant temperature for storage. The data provided by RFID tags will certainly be useful in tracking such perishable items [8].

4.7.1 Internet of Things Benefits in Retail and Supply Chain Management

- Increased supply chain transparency.
- Automated check-in and check-out of goods.
- Keeping an eye on the location of goods and warehouse storage conditions.
- Predictive equipment maintenance.
- Inventory management and theft prevention.

- Enhancing the shopping experience and providing superior customer service.
- Accurate and timely notification of any issues encountered during transportation.
- Demand alerts for warehouses and route optimisation. (htt3).

5 Iot in Manufacturing

5.1 IoT and Industry 4.0

Industrial systems' new definitions have recently included trendy terms such as Industry 4.0, the Internet of Things (IoT) and physical cybersystems. Additionally, whilst Industry 4.0 and related technologies offer numerous potential benefits for manufacturing, the vast majority of these innovations are implemented for and/or by large businesses [9]. The fourth industrial revolution is triggered by real-time network communication of products, processes and infrastructure. The Internet connects production, procurement, maintenance, delivery and customer service. Value chains that were once rigid are being transformed into highly adaptable value networks. The Internet connects production, procurement, maintenance, delivery and customer service. Value chains that are rigid are being transformed into highly flexible value networks. The Internet of Things provides an important model for businesses operating on a global scale to vertically integrate intelligent manufacturing machines, products and resources into flexible and vertically integrated manufacturing systems. It enables horizontal integration into the cross-industry value network. As a result, all countries view networking and digitisation as Industry 4.0's top priorities [10].

5.2 Usage of IoT

IoT is used in many fields as well as in many sectors related to production and manufacturing. It is frequently used in the production sector both alone and together with the concepts of cyber-physics. Remote monitoring, supply chain management and optimisation, digital twin and real-time machine monitoring, predictive maintenance, production support, production and production visibility and automation include service's IoT. This section is the basis on which concept it is used for production with its general framework.

5.2.1 Remote Monitoring

Wind turbines, for example are widely used in metallurgy, electricity, petroleum and chemical engineering, and they contribute significantly to industrial production.

By and large, various components of equipment are easily damaged as a result of complex working conditions and long work hours, resulting in a halt to production and significant economic losses. Stable operation and deterioration of equipment affect not only production capacity, product quality and costs but also production safety. Much effort has been expended to promote technological advancements. Certain critical equipment remains difficult to monitor due to the annoyance of wires but remains difficult to monitor due to the annoyance of wire transmission and power supply. With the rapid advancement of IoT, Industry 4.0 and smart manufacturing have been resurrected, further increasing the automation level of modern industrial production through monitoring and control [11, 12] (Wei et al.).

Vibration, temperature and current are all examples of signals that can be measured and transmitted to the machine owner via IoT and wireless sensor networks installed in critical equipment components. The server is then configured with a physical or data-driven model that performs anomaly detection, that is, the identification of abnormal activity states using sampled sensor data. When combined with failure causes, this enables the detection and elimination of errors in real time, as well as the identification of underlying failures in advance, thereby reducing safety risks and extending service life, prolong the life of equipment and reduce business costs [13].

5.2.2 Digital Twins

The concept of the digital twin is gaining traction in industry and, more recently, academia. Industry 4.0's development has been facilitated by advancements in its concepts, particularly in the manufacturing industry. Although the term "digital twin" is used loosely, it is best defined as the effortless integration of data between physical and virtual machines in both directions. The article discusses the challenges, applications and enabling technologies associated with artificial intelligence, the Internet of Things (IoT) and digital twins. The fourth industrial revolution enables the Digital Twin concept to become a reality for manufacturing processes by leveraging device connectivity. By utilising a Digital Twin, you can increase the connectivity and responsiveness of your devices, thereby increasing their reliability and performance. Due to the machine's ability to store large amounts of data required for performance analysis and prediction, AI algorithms combined with Digital Twins have the potential for increased accuracy. The automotive industry, particularly Tesla, is one application of Digital Twins. The ability to create a Digital Twin of an engine or automotive component can be advantageous when simulating and analysing data. AI improves test accuracy by performing data analysis on live vehicle data to forecast component efficiency in the present and future [14].

5.2.3 Real-Time Machine Monitoring

RFID and wireless communication technologies are at the heart of the Internet of Things, as they are used to monitor the state of physical machines. After being processed by various data models and cloud-based services via smartphones, this information is presented through a graphical dashboard. For the purpose of collecting machine data, IoT devices are deployed in a CMfg environment, such as a workshop. Machine tools, which are a critical component of CMfg's shared resources, require continuous monitoring. By maximising the use of IoT technology, it is possible to identify and track various production resources. Decision-making in IoT-enabled manufacturing is qualitatively distinct from traditional manufacturing. Manufacturing processes generate massive amounts of data when IoT devices and deployed sensors are used. Utilising this data to its full potential for production analytics such as quality analysis, performance evaluation and market forecasting is invaluable.

5.2.4 Predictive Maintenance

Businesses are now experimenting with various predictive maintenance techniques in order to reduce the cost and frequency of maintenance activities. Because IoT platforms can integrate data from multiple machines and production systems, they are ideal for predictive maintenance. The Internet of Things enables a new, simple method of monitoring and predictive maintenance for industrial equipment. It includes numerous features for connected manufacturing, such as process monitoring to ensure continuous quality and monitoring to avoid unplanned downtime [15].

5.2.5 Production Visibility

The Smart Factory is a critical component of Industry 4.0, which is the industrial revolution of the future. The Internet of Things (IoT) enables the Smart Factory Visibility and Traceability Platform to visualise manufacturing operations in real time in a smart factory. iVTP makes use of Internet of Things technology to specify various fabrication objects. More precisely, radiofrequency identification (RFID) devices are used to convert disparate resources into intelligent manufacturing objects, whose interactions can be reflected in real-time activities and production behaviour. By utilising an innovative laser scanner in the workshop, we can display the real-time movement of various smart manufacturing objects and match the RFID data to show their status. The cloud-based system configuration enables the packaging and deployment of all services in the cloud, enabling common end users to easily define their production logic, download useful services and develop custom services. Numerous case studies demonstrate how can be used to support typical decision-making, manufacturing, and logistics operations in a smart factory.

5.2.6 Automation

Modern products are in the trend smarter, more flexible and more sophisticated. The product structure becomes even more compound, which causes serious difficulties in the assembly process. System performance, such as profitability, delivery time, quality and cost, depends largely on the efficiency of the assembly model. The difficulty level of the assembly model depends on the complexity of the product and the availability of data for assembly planning. IoT and cloud computing in business systems will overcome the bottlenecks. The Internet of Things and cloud computing aim to help transform a traditional assembly model system into an advanced system capable of automatically handling complexity and change. To this end, a robust, reliable automated modelling and assembly system are proposed, establishing a system of flexible and extensible system components, integrated object models and interfaces to facilitate the following innovations: inclusive modular architecture and automated algorithms to get associative assembly matrices for assembly planning [16].

5.3   Manufacturing Applications of IoT

In this section, the applications of IoT and support technologies in the manufacturing industry are examined.

Machines and computers work together in Siemens's electronic manufacturing plant in Amberg, Germany. This is an example of the Internet technology of Things. Machines and computers are autonomous throughout the production line with 1000 automation controllers; 75% of the value chain product codes keeps the features, requirements, and information on the part, such as what actions it will see in the next step. Parts communicate with machines through these product codes. All processes are autonomous in information technologies control. This reduces the likelihood of errors in production processes. Employees are also responsible for overseeing any unexpected situations that may occur during this process. (htt6).

One of GE's battery factories has more than 10,000 sensors. These sensors are environmental conditions (temperature, humidity, air pressure, etc.) measured regularly. These sensors also collect the data generated as a result of operations on machines. In this way, manufacturers have the opportunity to observe battery performance in different conditions as they regularly measure environmental conditions, and they have access to real-time information about manufacturing and product processes. (htt6).

Additionally, GE wind farms employ sensors. Wind bleachers collect enormous amounts of operational data. Through digital monitoring of wind systems, GE uses this data to provide information to renewable energy users. They are concerned with mitigating the risks and costs associated with insulators. GE provides much more than wind turbine and wind farm data and insights. Their OEM-independent applications analyse, operationalise and monetise assets, enabling them to improve the management and performance of digital wind farms. (htt7).

BMW's production plant has intelligent energy metres based on IoT. These gauges regularly measure energy consumption to identify and predict sources with abnormally high energy expenditures. The first year of use of this technology has saved

100,000 kWh of energy, and with this savings, they are expected to provide a 25 million euro cost advantage over the next 10 years [17].

It is possible to see the use of IoT technology in conjunction with different technology applications like VR. Service technicians at the Thyysencroupp company use mixed reality devices in elevator service operations. Hands-free remote holographic guidance reduces the average length of service calls by up to four times. (htt8).

6 Challenges

Whilst the Internet of things is constantly evolving and growing technology, it seems to have a profound impact on our lives. The concept of the Internet of things is thought to evolve as the Internet of Everything, but this dream will not be so easy to reach; there will be some challenges encountered. Let's look at these challenges one by one [18].

6.1 Socio-Ethical Considerations

Technological developments affect society directly or indirectly. As well as being useful to society, the developed product must be acceptable to society. It is possible that non-socially adopted technological products will not create a demand in the market, so they will not see the expected value.

We saw a similar example of this in an example of glasses developed by a famous brand. He was not accepted by society because he thought it was damaging personal privacy, and he was subjected to restrictions by the government. In this case, the brand directed the product into industrial use instead of personal use.

Another issue that should be addressed is social concerns. Developed products can connect with other objects in different ways, making this product vulnerable to attack. So, users are concerned about the capture of their data, security issues.

Many people find it difficult to accept the technology of the Internet of Things, thinking that the security measures they expect are not met. This is an obstacle to achieving the dream of the Internet of Everything, as the adaptation of societies is slow to develop [6].

6.1.1 Scalability

Scalability is an important concept in the use of Internet technology of Things, especially in the industry. Scalability means that the current system can also operate effectively when the number of users, capacity or production volume increases. A scalable system continues to work effectively even when the number of resources and the number of users are increased significantly. Businesses expect the technology they add to their systems to be scalable at a high cost. The development of this technology requires low power, continuous data flow through the uninterrupted network. The Internet solutions of existing objects can be inadequate to meet what is expected [7].

These sorts of systems require good performance, have to achieve reliability and support scale [6].

6.2   Technological Limits

6.2.1   Radio Spectrum

IoT is set to enable a large number of unconnected devices to communicate and share data with one sort another is, and the wireless world communication leads to the need for more spectrums. In the next decade, connecting 50 million devices to the Internet would require more spectrum than is available today [6].

6.2.2   Battery Life of Future IoT Hardware Platforms

The amount of battery energy consumed by an IoT device is highly dependent on the radio transmitter type, the communication protocols used, the sensors and the processor type. Lithium, nickel and alkaline batteries are all types of batteries that are frequently used in small IoT devices. The majority of these battery chemistries have extremely low self-discharge characteristics, which makes them ideal for extended service intervals. Nonetheless, battery life is a critical factor to consider when designing and deploying IoT devices for specific applications. This is because replacing batteries in the field is not economically feasible, even more so when thousands or even millions of devices are involved. Developing ultra-low-power consumption devices and extending the battery life of Internet of Things devices are active research areas [18].

6.2.3   Failure Handling

Another characteristic of Internet of things technology that is expected is fault tolerance. The term "failure handling" refers to the processes of detecting, masking and tolerating failures, as well as recovering from failures and redundancy. In a complex, dynamic and heterogeneous environment as the Internet of Things is expected to be, systems must be capable of self-configuration, self-diagnosis and self-repair.

In particular, technologies used in the industry are expected to have fault tolerance. Businesses will not want to include technology without fault tolerance in their systems. They want systems to have the ability to anticipate mistakes, repair themselves, thinking that a minor error in the system could cause serious material and moral losses. The development of this technology, which has many features, is difficult for developers [18].

6.2.4   IoT Data and Information Processing

The Internet of Things technology collects large amounts of data through sensors. Where, in what ways, where these data are collected, and how the collected data are processed is an important research issue. It is a matter of much research is done today that can generate value from this collected data, be able to access the right data at the right time to benefit the individual or institution in the decision-making phase and be able to process this data. So, with wisdom and proper knowledge, extraction of real information from sensor data is very useful and desirable [18].

Due to the heterogeneous nature of the data collected, as well as its volume and complexity, effective IoT data usage presents challenges [7].

Interoperability.

Heterogeneity has been a significant challenge in distributed systems since a variety of networks, hardware, operating systems and programming languages began coexisting within the same system.

Users, machines and objects are all connected via the Internet of Things. From this vantage point, this technology can be utilised by all of these entities cooperating. It is referred to as interoperability in the literature. Despite the fact that the IoT faces additional challenges, interoperability remains one of the most difficult goals for IoT systems unless a set of standards is developed [7].

6.3   IoT Security

The Internet of Things is a multibillion-dollar market in the consumer, enterprise and industrial sectors. Their application areas are expanding daily, implying that they will eventually have a greater impact on our lives. Any system that contains the Internet or any of its variants is attackable. Although various standards and protocols have been established to ensure that these systems meet the necessary security requirements, there are currently no complete solutions to security problems. Institutions and individuals are hesitant to incorporate new technologies into their daily lives out of concern for the possibility of unauthorised access to private information. This represents a significant impediment to the development of Internet of Things technologies. Each component of the system is subjected to a unique set of security concerns [18].

In addition to the already complex security and privacy landscape, the Internet of Things introduces a slew of new data security and privacy concerns. Frequently, IoT systems rely on wireless communications, which by definition pose security risks. Additionally, the large volume of data generated raises new concerns about not only managing, processing and analysing such a large volume of data but also about ensuring data confidentiality. IoT systems, particularly those that collect sensitive data (for example healthcare systems), must be secured at all layers, from the physical to the application. Existing Internet of Things-enabled devices and deployed systems are especially vulnerable to denial-of-service attacks. Only by implementing adequate security and data protection mechanisms can IoT systems earn the users' trust. Security and privacy concerns should be addressed from the start of the system's design [7].

6.4   Legislation and Governance

The interaction of technology and regulation, or government, has historically been fraught. Most governments are hesitant to legislate on technology issues, claiming that legislation stifles innovation, preferring technology-neutral regulation that does not require changes with each new advancement. The requirement for infrastructure and communication protocol standardisation, the movement of data across national borders and the associated cloud-based databases will necessitate some form of governance, even if only to ensure the IoT infrastructure functions properly. Additionally, the need to protect individual privacy is frequently outweighed by the need for organisations to use data collected from IoT systems. Although governance of IoT systems is rarely discussed in the literature, the necessity of resolving issues such as integrating disparate IoT systems owned by competing organisations, adhering to national regulation, ensuring individual privacy and the confidentiality of business data and securing IoT infrastructure against attack is likely to require significant research and organisational resources. Nobody can predict the unintended benefits or consequences of the IoT, and the system's management is ambiguous. Given the large number of interested parties, a multistakeholder governance model is the most likely. It is highly improbable that a single organisation or governmental entity will be able to control the IoT [7].

Environment.

In the coming years, the Internet of Things will create a mass environment replete with sensors, devices and objects. On the one hand, it encompasses strategies for reducing carbon emissions and pollution whilst increasing energy efficiency; on the other hand, it is a method for developing energy-efficient computing devices, communication protocols and networking approaches for interconnecting devices. In the literature, this Internet of Things concept for energy efficiency is referred to as Green IoT. The Green IoT can be integrated with solar panels powered by technology. Additional work in Green IoT is required in the future. The IoT solutions should provide feedback on energy consumption and should direct the user towards more efficient energy allocation. Energy harvesting should be implemented in the environment. Energy-related issues such as energy harvesting and low-power chipsets are critical for the development of the Internet of Things. Current technology support is insufficient for processing power, and energy consumption will soon become a problem. Although significant efforts have been made to advance IoT technologies, Green IoT continues to face numerous significant challenges in a variety of areas. Additional research in this area is necessary [6].

Internet of Things technologies are predicted to be as detrimental to the environment as they are beneficial. Our technological devices have a limited life and lifespan. When the usage time period expires, these devices become inoperable. These devices become obsolete and are disposed of electronically into nature. Certain components of these devices are recyclable. However, they contain substances that can be harmful to the environment and other living things. These substances are absorbed into the soil or air, posing health and environmental concerns. Electronic products contain a number of hazardous substances that are known to be harmful to human health, including risks to the brain, nervous system, lungs and kidneys, as well as links to certain cancers. Toxic residues can leak into and contaminate the soil, air and water, affecting the ecosystems in which indigenous people grow food, hunt and fish. Hazardous substances are also transported between continents via the air and sea. Around 50 million metric tonnes of electronic products are discarded each year, according to the United Nations University's Global e-waste monitor. E-waste is the world's fastest growing waste stream, expected to reach 52.2 million metric tonnes by 2021 unless the trend is reversed. Numerous factors contribute to the increase. The world's population is increasing, and economic prosperity is reaching an increasing number of people. Technological advancements are accelerating, and the cost of information technology products is decreasing, resulting in shorter product lifespans. Whilst developed countries continue to contribute the majority of the problem, developing countries are catching up rapidly. There has been insufficient research on the e-waste problem to date, and no solution has been discovered. Our future will be jeopardised if the states do not resolve this issue. Most countries struggle to manage these massive amounts of discarded products responsibly and efficiently. Globally, only, 20% of electronic waste was recycled in 2016. Whilst the emphasis has been on product collection, insufficient effort has been made to develop infrastructure for waste processing and safe material recovery. This has resulted in a scarcity of facilities capable of safely managing e-waste. Exporters of e-waste frequently choose destinations that lack effective legislation governing how e-waste should be handled. (htt4).

7 The Future of Iot

The Internet of Things (IoT) has become a popular research topic in recent years because it enables various items to communicate without human intervention. The Internet of Things (IoT) transforms physical objects into intelligent objects through the use of enabling technologies such as Internet protocols, wireless sensor networks (WSNs), pervasive and ubiquitous computing, applications and communication technologies. Aggregation, compilation, processing, analysis and mining of data are all required steps in acquiring critical information for ubiquitous and intelligent services.

The Internet of Everything (IoET), the Internet of Vehicle Things (IoVT), the Internet of Social Things (IoST), Sensor as a Service (SaaS) and the Fog of Everything (FoE) have all been added to the IoT concept in recent years. These game-changing concepts can assist in determining which core technologies, communication protocols, and data privacy policies should be used in an IoT ecosystem to accomplish any given goal [19].

7.1   Internet of Everything (Ioe)

The Internet of Everything (IOE) connects people, processes, data and things, transforming data into actions that enable new capabilities, richer experiences and unprecedented economic opportunity for businesses, individuals and countries.

In simple terms, the Internet of Everything is the network of connections between people, things, data and processes that form a unified, interconnected system with the goal of enhancing user experiences and enabling smarter decision-making.

The Internet of Everything (IOE) envisions a world in which billions of items are connected via public or private networks and monitored, measured and assessed using standard and proprietary protocols [20].

The Internet of Everything (IOE) ideology envisions a society in which billions of sensors are embedded in billions of devices, machines and everyday things, allowing them to connect and become smarter.

The Internet of Things' core purpose is to turn collected data into actions, make data-driven decision-making easier and deliver new capabilities and richer experiences. (htt1).

The IoE concept originated with the idea of enabling automated machines via ubiquitous Internet, large data processing, and artificial intelligence. According to Cisco, the Internet of Things is built on the "four pillars" of people, data, process and things. This perspective implies that the Internet of Things (IoE) includes not only "things" but also "automated and human-centred processes" (i.e. intelligent machines/devices).

This concept extends far beyond the IoT's category of connected "things" (i.e. pure machines/devices). Additionally, the advancement of big data and artificial intelligence technologies adds to the IoE's building blocks. Recent literature has reintroduced the essence of IoE, namely accumulating massive data hidden from the Internet using various AI algorithms and providing automated capabilities to all devices/machines. As a result, the Internet of Things is capable of harvesting and analysing real-time data from millions of connected devices, as well as taking proactive, intelligent decisions, enabling "automated intelligence."

Whilst the concept of the "Internet of Everything" has been discussed and debated for years, its implementation is still in its infancy. Regardless of the challenges associated with fully realising IoE, the alluring vision of IoE will never dissuade us from adopting it [21].

The Internet of Everything (IoE) is a concept used in information technology to describe a device that combines sensing, processing, information extraction and communication capabilities. The Internet of Things (IoE) enables various electronic devices with varying capacities to detect the environment and communicate for data exchange. The Internet of Things (IoE) is a generic term for wireless sensor networks. Different classifications, types and capabilities of IoE nodes exist. Smartphones, tablets, laptops, home appliances and even automobiles are examples of nodes in the Internet of Things. These nodes can use their many sensors to detect the environment, interpret data, acquire important information, communicate over the Internet and adapt their behaviour. The ability to communicate and exchange information determines the smartness and intelligence of IoE nodes, not their computational capacity. These gadgets can learn from their observed data thanks to communication linkages. It teaches these devices how to use their data to do new and productive tasks [22].

IoE has the potential to significantly contribute to the creation of new prospects for Industry 5.0 applications. Industry 5.0 advances by the Internet of Things can give new functions, a better user experience and expected benefits for industries and governments. The Internet of Things' involvement in Industry 5.0, including increasing consumer loyalty and enjoyment and customising experiences based on data supplied by IoE. The usage of IoE in Industry 5.0 reduces latency and eliminates bottlenecks on communication routes, resulting in lower operating expenses [23].

7.1.1 IoE Features

- Decentralisation and edge computing—rather than processing data in a single hub, data are processed across multiple distributed nodes.
- External data can be ingested and then returned to other network components.
- Relationship with each technology involved in the digital transformation process—cloud computing, fog computing, artificial intelligence, machine learning, the Internet of Things and big data, to name a few. Indeed, the growth of big data and the advancement of IoE technologies are inextricably linked.. (htt1).

7.1.2 Pillars of the Internet of Everything (IoE)

The Internet of Everything (IoE) is the intelligent linkage of four pillars:

- People
- Processes
- Data
- Things.

To take advantage of the Internet of Things, all of the things in these four categories should have sensors that can detect, measure, assess and generate data.

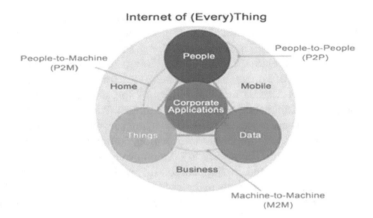

Fig. 4 Internet of everything (IOE) [25]

Let's take a closer look at the four pillars of IoE and how they interact to create IoE:

1. **People**: humans can be equipped with sensors in many ways—consider smart-watches for a less alarming example—that can generate data.
2. **Data**: data will be converted into intelligence that can be used.
3. **Processes**: getting the right information to the right recipient on time, whether it is a human, an algorithm or a machine.
4. **Things**: the Internet of Things allows physical devices and objects to communicate with one another (IoT) [24] (Fig. 4).

7.1.3   The Difference Between IoE and IoT

Numerous readers frequently struggle to understand the terms Internet of Things (IoT) and Internet of Everything (IoE). Nano-things and the Internet of Everything (IoE): whilst the IoT is composed entirely of objects, the IoE is founded on four pillars (data, people, process and things). The Internet of Everything is capable of retrieving real-time data from millions and billions of connected sensors. Cisco defines the Internet of Things (IoE) as an intelligent connection of processes, people, information and objects. The Internet of Things' objective is to unite physical and virtual objects into a single entity. It is about enabling an entire ecosystem of living, non-living or virtual services or products to communicate with one another, not just smart gadgets transmitting signals or communicating via some channels. This digital component is absent from the Internet of Things concept. Whilst the Internet of Things encompasses smart devices (connected to a variety of things and people) and the availability of an Internet connection, it lacks a digital component (sort of a digital object practically equivalent to any real-world object). The Internet of Things' network of connections may be physical–physical, human–human, digital, human–physical,

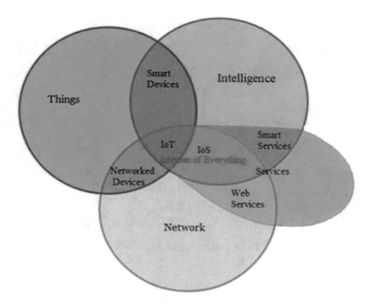

Fig. 5 Representation of internet of everything via venn diagram [26]

physical–digital or human–digital. Nevertheless, due to certain distinguishing characteristics, such as digital presence, both of these concepts are quite similar yet slightly dissimilar. Figure 5 illustrates these explanations as a Venn diagram to aid in topic comprehension [26].

7.1.4 Three Expectations of IoE

The Internet of Everything's (IoE) goal is to connect ubiquitous electronic devices (i.e. IoE terminal nodes) to the Internet, analyse the massive amounts of data generated by connected terminal nodes and then offer intelligent applications to help human society grow. To realise this vision, the IoE is expected to meet three primary criteria:

- Scalability refers to the ability to create a scalable network architecture with global coverage;
- Intelligence refers to the ability to create a global computing facility capable of making intelligent decisions,
- And diversity refers to the ability to handle a diverse set of applications. In Fig. 6, three expectations are depicted alongside their typical enabling technology. The three expectations are detailed below in their entirety.

1. Scalability: scalability refers to the ability of an IoE network to cover everywhere and everything elastically. Thus, IoE can be used to address a variety of

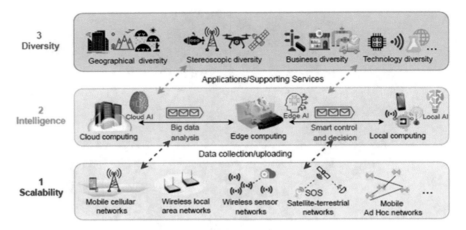

Fig. 6 Three expectations of IOE (i.e. scalability, intelligence and diversity) [21]

communication requirements in a variety of environments, including urban, rural, aquatic, terrestrial, aerial and space. To accomplish this, the scalable IoE network must provide extensive coverage, massive access, and ubiquitous connectivity. Multiple communication technologies with varying transmission distances (from a few metres to a thousand metres) and topologies can be used to build such IoE networks.

2. Intelligence: in addition to enabling distributed computing across the IoE, intelligence entails enabling intelligent analysis, forecasting, decision-making and action across all devices. IoE, in particular, must collect massive amounts of data from its vast and scalable network, extract critical information from that data (such as intelligent instructions or decisions) and then use that information to enable intelligent actions or controls across the board.

3. Diversification: diversification refers to the breadth of applications that support the IoE's "automated and human-centred processes." Scalability and intelligence in IoE are required for computing capability, security, energy efficiency and network performance and are thus critical for the successful implementation of a large number of IoE applications [21].

7.1.5   Internet of Everything Examples

Practically, every company may benefit from incorporating the Internet of Everything approach into their procedures. Here are a few samples to get you started:

- Smart water and electricity metres can be installed by municipal systems for individuals and businesses to monitor usage rates and make cost-cutting and cost-cutting decisions.

- Sensors for predictive maintenance can be included into manufacturing to monitor equipment parts that need to be repaired or replaced. This cuts down on downtime and lowers repair expenses.
- Sensors and smart gadgets can be installed on vehicles by logistics and delivery businesses to improve delivery conditions and feasible routes. Companies can eventually increase end-user happiness. (htt1)
- Nike unveiled an auto-lacing shoe in 2019 that can monitor the wearer's blood pressure using embedded sensors. It will then loosen or tighten the shoe on its own, as needed.
- Tel-municipality Aviv's installed Cisco's smart-water camera chips in water pipelines. The supervisory control centre was able to monitor leaks, drains, water levels and pressure by sending data from subterranean pipes to the cloud. Water-smart technology has the potential to lower costs, avert water shortages and improve user efficiency.
- Sensors used in machinery and equipment in the industrial business can correctly calculate physical coercion and monetary capital degradation. Preemptive maintenance, comparable to an odometre on a car, would be possible, decreasing erosion and downtime. It may, on the other hand, correctly evaluate its economic life and alert the corporation to the best time to offload the equipment.
- AI is already able to learn about a customer's preferences and recommend similar products in the online retail industry. However, the Internet of Things has the potential to go even further, albeit with data security concerns [24].

7.1.6   IoE's Pitfalls

Finally, we will look at three IoE pitfalls:

1. Security: this is, in many ways, an insurmountable IoE difficulty. The threat of cybersecurity exists as long as data are stored in the cloud. For example, if a family utilises a smart lock for entry and exit and the programme is hacked, hackers may hypothetically lock the family out of their own home.
2. Privacy: privacy is a major concern for IoE because it is linked to security. Take, for example smartwatches. They are already capable of forecasting a heart attack based on vital sign detection. If a health insurance provider has access to your medical data via your smartwatch, they will be able to discriminate against you in the form of an expensive blanket policy.
3. Compatibility: the number of Internet of Things (IoT) devices is increasing. Each year, thousands of different programmers use diverse methodologies, codes and standards to create additional gadgets and applications. Connecting all of these devices into a unified ecosystem may necessitate increased management, higher IT costs and lengthier downtime in the event of repairs [24].

7.2 Sensor as a Service

The Internet of Things (IoT) is a contemporary trending technology in which multiple sensors are connected to the Internet in order to manage assets efficiently and attractively in a Smart Environment (viz. Smart home, Smart city, etc.). Cloud technologies are used to deliver software, infrastructure and platforms as a service. Sensing as a Service (SaS) is a form of advanced distributed computing that leverages the Internet of Things' global resources and enables the establishment of a shared network of sensors that can be consumed as a service. Organisations and developers can leverage these services to enable consumers to monetize their data through the use of existing infrastructure.

Sensors embedded in smart devices are used to perform IoT-related tasks. These sensors gather data as they traverse their deployment area. The required data are received by data collectors or sensors via a pay-as-you-go IoT application, with data distribution and collection managed in the cloud. Real-time data from sensor devices are monetised and transformed into new data streams in order to improve the accuracy of predictive services and optimise network operations for effective automation [27].

Sensors are capable of tackling the problems faced by Smart Cities. Many sensors will be available as a result of getting 50 billion devices connected to the Internet by 2020. Many commonplace objects are still equipped with sensors; however, their use is limited to the object itself. Four conceptual levels make up the sensing as a service model:

(1)   Sensors and sensor owners,
(2)   Sensor publishers,
(3)   Extended service providers and,
(4)   Sensor data consumers.

Sensors and Sensor Owners Layer: this layer is comprised of sensors and sensor owners. A sensor is an electronic device that detects, measures or senses physical phenomena such as humidity, temperature or other environmental variables. Multiple sensors can be attached to an object or device. Today, there are numerous types of sensors available. They are capable of measuring a wide variety of occurrences. Additionally, they are capable of transmitting sensor data to the cloud [28].

Sensing as a Service is a novel concept that will be built on top of existing Internet of Things and cloud computing infrastructure. Sensing as a Service is based on the principle of providing client applications with Internet access to sensors that are managed and deployed by other entities. Sensor data are made available to data consumers for free or at a cost to sensor owners via the Internet. Sensing as a Service is viewed as a component that sits between two Internet of Things data sources and Internet of Things applications across a variety of domains, including smart cities, agriculture, manufacturing and health care, to name a few. Cloud-based Sensing as a Service middleware technologies are expected to play a critical role in delivering sensor data to IoT applications [29].

7.2.1   SaS MODEL

The cloud platform and worker nodes are the two most important components of SaS architecture. The platform manages the overall sensing activities, whilst the worker nodes carry out the tasks according to the cloud IoT platform's specifications.

Figure 7 depicts the generic design of a sensing model. The use of numerous sensing servers allows for the management of sensing requests from various places. A cloud user initiates a sensing request from a computer or a sensor device via a Web-based application, which is then delivered to the needed destination. Data collected from IoT devices are saved in the cloud and distributed to the appropriate users. The sensing chores are on par with the sensing requirements of other sensor users, which are an important characteristic of this functionality [27].

7.2.2   Architectural Requirements of the SaS Model

1. The model should provide a layered system capable of handling environmental conditions and network channels whilst consuming the least amount of power possible.
2. To carry out the fundamental operations across a large range of heterogeneous sensor devices, essential abstractions must be supplied.
3. It is necessary to provide a consistent mechanism of integrating higher-level systems in order to share sensor data.
4. A good model should be capable of managing both dynamic and static networks [27].

7.2.3   Benefits and Advantages of the Sensing Model

1. Integrated cloud computing: The sensing model can inherit the characteristics of popular cloud computing models such as IaaS, PaaS and SaaS.
2. Participatory Sensing and Actuation: Sensors are rapidly deployed across large geographic areas, effectively distributing workload and reducing the effort required to capture various events.
3. Reuse and sharing: The sensing model suppresses the characteristics that promote sensor data distribution and sharing.
4. Data monetisation: By repurposing existing network assets to create new data streams, real-time data about an area can be commercialised [27].

7.2.4   Sensing as a Service Applications

As a Service Sensing Middleware enables the development of IoT applications across a range of industries. This section discusses IoT applications that can be enabled by Sensing as a Service Middleware.

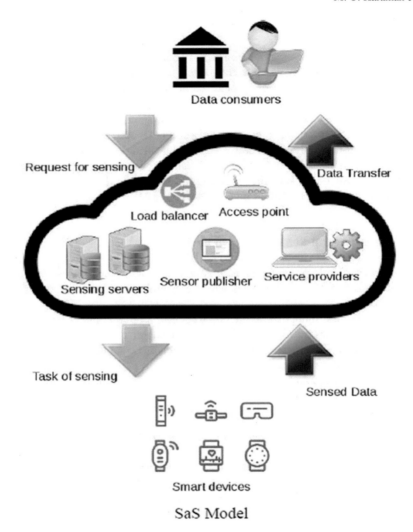

Fig. 7 SaS model [27]

(1) Sensing as a Service: Tracking and monitoring via the Internet Middleware can be used to keep track of remotely located objects of interest. As a result, the middleware can be used to generate real-time warnings and respond to user actions. Remote tracking and monitoring applications include environmental conditions, animal behaviour, automobiles, patient health conditions, building surveillance and security, vegetation production quality and smart-grid operations.

(2) Real-time resource management via Sensing as a Service Middleware can be used to control and optimise resources in real time, resulting in cost savings and

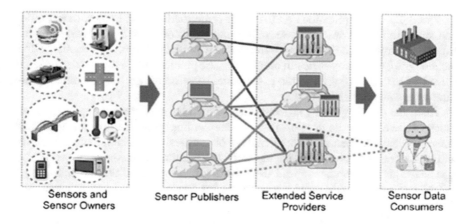

Sensing as a Service Model for IoT applications

Fig. 8 Sensing as a service model for IoT applications [29]

system efficiency improvements. The middleware may enable applications in a variety of fields, including guided navigation, traffic control, intelligent parking, waste management and water/irrigation management, all of which require real-time resource management.

(3) Middleware that provides Sensing as a Service enables remote fault detection in a variety of sectors, including network systems, automotive, aviation and aerospace, smart grids and oil and gas pipelines [29] (Fig. 8).

8 Conclusion

In this chapter, the Internet of Things technology and its applications in the field of manufacturing are introduced. The definition of IoT technology, its architecture, common applications in daily life and applications in the field of manufacturing are represented. In addition, the sensitivities and challenges in the use of this technology are examined, and technologies such as Service as a Sensor and Internet of Everything, which will be heard more frequently in the near future, are mentioned.

Literature shows that IoT applications do not have a uniform architecture. Each application creates its own structure. But, IoT applications generally have a three-layer structure as sensing, network and cloud computing. In this section, these layers and their properties are elucidated. In particular, it has been determined how important the RFID system is for IoT technology. For this reason, it has been observed that RFID systems are very efficient and at the same time inexpensive in the sensing and network stages.

The daily life applications of IoT are exemplified. Smart homes, smart cities and smart farms are described. In addition, it has been explained how people can follow their own data thanks to wearable smart products. Also, the benefits of IoT technologies, especially in health monitoring, are pointed out. In this sense, the rapid spread of IoT technology and the exponential increase of devices connected to the Internet are emphasised.

Then, the challenges of IoT technology, which has penetrated people's lives so much and will show its deeper effects in the coming years, are examined. In this context, considerations under the headings of socio-ethical, technological, security and legislation and governance are analysed. The fact that this technology has entered human life at an unstoppable speed also reveals some responsibilities. For this reason, IoT technologies should provide a more secure service under the titles mentioned above.

It is considered that with the more widespread use of IoT technologies in the future, more complex IoT devices will affect human life. Data, process, people and thing can be combined under a single definition, and millions of data can be evaluated at the same time. This is called the Internet of Everything. In this way, all infrastructure social services of a city can be controlled through smart applications. Almost all the processes and needs of a production line can be followed in the field of manufacturing. In addition, Internet of Things is not affected by the problems related to scalability whilst doing this. In this section, examples of Internet of Everything and its differences from Internet of Thing are mentioned with the above concept. Then, Sensor as a Service technology and architecture are mentioned. Finally, Sensor as a Service applications are also exemplified.

References

1. Xia F, Yang LT, Wang L, Vinel A (2012) Internet of things. Int J Commun Syst
2. Nord JH, Koohang A, Paliszkiewicz J (2019) The internet of things: review and theoretical framework. Expert Syst Appl 133:97–108. ISSN 0957-4174
3. Trujillo-Rasua R, Solanas A, Pérez-Martínez PA, Domingo-Ferrer J (2012) Predictive protocol for the scalable identification of RFID tags through collaborative readers. Comput Ind 557–573
4. Yuan YS, Zhang J (2013) Development of wireless sensor network based on ZigBee and RFID technology. Appl Mech Mater 1175–1180
5. Lu D, Teng Q (2012) A application of cloud computing and IOT in logistics. J Softw Eng Appl 204–207
6. Anuradha J, Tripathy BK (2018) Internet of things (IoT) technologies, applications, challenges and solutions. CRC Press_Taylor & Francis
7. Hassan QF, Madani SA (2017) Internet of things: Challenges, advances, and applications. CRC Press
8. Agarwal S, Makkar S, Tran DT (2020) Privacy vulnerabilities and data security challenges in the IoT. CRC Press
9. Saravanan GE (2021) Implementation of IoT in production and manufacturing: an industry 4.0 approach. Mater Today Proc
10. Henning Kagermann RA (2016) Industrie 4.0 in a global context. Herbert Utz Verlag

11. Korkua SJH (2010) Wireless health monitoring system for vibration detection of induction motors. In: 2010 IEEE industrial and commercial power systems technical conference. IEEE, pp 1–6
12. Xu LD, He W, Li S (2014) Internet of things in industries: a survey. IEEE Trans Ind Inform 2233–2243
13. Lia C, Moa L, Tanga H, Yanb R (2020) Lifelong condition monitoring based on NB-IoT for anomaly detection of machinery equipmenT. Procedia Manuf 144–149
14. Fuller A, Fan Z, Day C, Barlow C (2020) Digital twin: Enabling technologies, challenges and open research. IEEE, pp 108952–108971
15. Parpala RC (2017) Application of IoT concept on predictive maintenance of industrial equipment. In: 8th international conference on manufacturing science and education—MSE 2017 "Trends in new industrial revolution" (s. 8). MATEC web of conferences
16. Wang CZ (2014) IoT and cloud computing in automation of assembly modeling systems. IEEE Trans Ind Inf 1426–1434
17. Lödding H (2017) Advances in production management systems. The path to intelligent, collaborative and sustainable manufacturing. Springer, Hamburg
18. James A, Seth A, Mukhopadhyay AC (2021) Smart sensors, measurement, and instrumentation. In: IoT system Design_Project-based approach
19. Pourghebleh B, Hayyolalam V (2020) A comprehensive and systematic review of the load balancing mechanisms in the Internet of Things. Cluster Comput 641–661
20. Vinothini V (2020) Big data with IOT (Internet of Things), IOE (Internet of Everythings): a review. Int J Res Appl Sci Eng Technol (IJRASET)
21. Liu Y, Dai HN, Wang Q, Shukla MK, Imran M (2020) Unmanned aerial vehicle for internet of everything: opportunities and challenges. Comput Communi 66–83
22. Masoud M, Jaradat Y, Manasrah A, Jannoud I (2019) Sensors of smart devices in the Internet of Everything (IoE) era: big opportunities and massive doubts. J Sens
23. Maddikunta PK, Pham QV, Prabadevi B, Deepa N, Dev K, Gadekallu TR, Liyanage M (2021) Industry 5.0: a survey on enabling technologies and potential applications. J Ind Inf Integr 100257
24. Ghosh A, Chakraborty D, Law A (2018) Artificial intelligence in internet of things. CAAI Trans Intell Technol 208–218
25. https://www.sam-solutions.com/blog/what-is-internet-of-everything-ioe/
26. Manavalan M (2020) Intersection of artificial intelligence, machine learning, and internet of things–an economic overview. Glob Discl Econ Bus 119–128
27. YR SK, Champa HN (2019) An extensive review on sensing as a service paradigm in IoT: architecture, research challenges, lessons learned and future directions. Int J Appl Eng Res 1220–1243
28. Perera C, Zaslavsky A, Christen P, Georgakopoulos D (2014) Sensing as a service model for smart cities supported by internet of things. Trans Emerg Telecommun Technol 81–93
29. Alarbi M (2017) Middleware architecture for sensing as a service
30. Saadati Z, Zeki CP, Vatankhah Barenji R (2021) On the development of blockchain-based learning management system as a metacognitive tool to support self-regulation learning in online higher education. Interact Learn Environ 1–24
31. https://easternpeak.com/blog/6-cool-examples-of-internet-of-things-applications-and-how-to-develop-one/
32. https://tcocertified.com/e-waste/
33. Barenji AV, Barenji RV, Roudi D, Hashemipour M (2017) A dynamic multi-agent-based scheduling approach for SMEs. Int J Adv Manufac Technol 89(9):3123–3137
34. https://www.ge.com/renewableenergy/wind-energy/onshore-wind/digital-wind-farm
35. Barenji RV, Barenji AV, Hashemipour M (2014) A multi-agent RFID-enabled distributed control system for a flexible manufacturing shop. Int J Adv Manufac Technol 71(9):1773–1791
36. Wei X, Hou L, Hao J (tarih yok) (2018) Machine fault diagnosis using IIoT, IWSNs, HHT, and SVM. In: 2018 IEEE 18th international conference on communication technology (ICCT). IEEE.

37. Zhonga Y, Wangb LR, Xua X (2017a) An IoT-enabled real-time machine status monitoring approach for cloud manufacturin. Procedia CIRP 709–714
38. Zhonga Y, Xua RX, Wangb L (2017b) IoT-enabled smart factory visibility and traceability using laser-scanners. Procedia Manufact 1–14

Blockchain Technology and Its Role in Industry 4.0

Ozge Yesilyurt, Kubra Nur Ozcan, Merve Melis Ergün, and Hatef Javadi

1   Introduction to Blockchain Technology

Industry 4.0 is the next phase in the transition of traditional factories into smart factories that are resource efficient and exceptionally adaptable to changing customer requirements. It is worth noting that the Industry 4.0 paradigm encourages the use of such technologies to enable the evolution of factory communications architectures away from cloud- or Internet-service-centric architectures and towards architectures in which all entities involved in industrial processes share information via a peer-to-peer (P2P) network. Blockchain technology, which evolved from the cryptocurrency Bitcoin and enables the development of decentralised applications (DApps) capable of tracking and storing transactions made by a large number of users and devices concurrently, is one of the most promising industrial applications technologies [1]. Within a decade, the blockchain has developed from a relatively obscure and poorly understood concept to a phenomenon that has garnered widespread media interest and drew academics and practitioners alike [2]. Blockchain technology is being heralded as one of the most transformative inventions of the previous decade, with the potential to disrupt practically every industry, from banking to manufacturing to education [3]. Blockchain technology has grown in popularity in recent years because to its decentralised, peer-to-peer transaction, and unchangeable qualities. It is a distributed ledger system that is open to all network users. The concept is based on Satoshi Nakamoto's 2008 Bitcoin cryptocurrency [4].

Due to the fact that Bitcoin was the first blockchain application, it is credited with coining the word "blockchain." Bitcoin's concept is based on the whitepaper "Bitcoin: a peer-to-peer electronic cash system," which was released in 2008 under the pseudonym Satoshi Nakamoto by an individual (or group). Bitcoin was founded

O. Yesilyurt (✉) · K. N. Ozcan · M. M. Ergün · H. Javadi
Department of Industrial Engineering, Hacettepe University, Beytepe Campus 06800, Ankara, Turkey
e-mail: contact.ozgeyesilyurt@gmail.com

in the aftermath of the 2008 global financial crisis, which was a crucial factor in its development. The technology gained widespread recognition in 2009, with the development of the Bitcoin blockchain network. Despite being designed as a decentralised alternative to the existing controlled financial currency system, Bitcoin was the first of several blockchain applications. Whilst blockchain technology is still in its infancy, the techniques used in blockchain date back several decades. In 1991, two research scientists, Stuart Haber and W. Scott Stornetta, envisioned blockchain technology [5]. They pioneered the use of time-stamped digital documents to prevent them from being backdated or edited [6].

BCT is a distributed open ledger that enables the recording and sharing of information amongst a group of persons. Each participant has a copy of the data, and any modifications must be approved by the group members. The data might represent a transaction, a contract, an asset, a person's identity, or nearly anything else that can be defined digitally. BCT entries are permanent, accessible, and searchable, enabling community members to view detailed transaction histories. Each pair of transactions results in the addition of a new "block" to the end of a "chain." A protocol specifies the steps involved in initiating, evaluating, recording, and disseminating new adjustments or additions. Cryptographic validation is performed to connect the records linked in the blocks [7].

Bitcoin was initially published in academic journals in 2012, followed by articles on the blockchain and distributed ledger technology (DLT) in 2015. This lag is partly due to the lengthy review cycles of prestigious academic publications, but it is also owing to the complexity of the technology, which is exacerbated by poorly understood and confusing application cases. This situation has improved as a result of the proliferation of publications aimed at a broader audience in which authors speculate on potential application scenarios for the technology. Numerous proposed use cases are broad in nature and span a range of sectors and applications, including financial services, transportation and supply chain management, media and entertainment, education, tourism, government services, consumer services, voting, and academic peer assessments [3]. This chapter will discuss the blockchain architecture, its operating principles, and its numerous varieties, as well as the benefits and drawbacks of its use, as well as its application potential in a variety of sectors.

2 Blockchain Architecture

In the field of information technology, the new word "blockchain" is self-explanatory: a "block" is a storage unit or space for digital information to be stored in a database, and a "chain" refers to the different or string of database locations where these blocks of information are stored.

However, what data is contained in each of these blocks? How long may a chain be?

Blocks contain specific information about digital/online transactions. Each block provides the amount of a transaction, the time, and date of the transaction. Rather of disclosing real names, the block stores the data (but not necessarily the identities) of transaction participants such as persons, businesses, and legal organisations by assigning a cryptographic hash or unique digital signature [6]. Blocks are then added to the chain once they have been validated by all participants using one of the several possible consensus techniques (note: the chain's very first block is referred to as the "genesis" block) [7]. New blocks are generated in accordance with predefined criteria and, in certain cases, a background activity dubbed "mining".

Blocks also interact with business rules governing the transfer and status of particular information inside each block, such as who owns what and when and between whom a transfer may occur. These rules, together with their implementation, are known to as "smart contracts," which are automated procedures built into the blockchain's operations.

Because blockchains are decentralised ledgers, there is no centralised point where they may be downloaded. New participants in a blockchain must install appropriate software on their laptop (for example, Web3py, a Python library for connecting to the Ethereum blockchain network) and then connect to the blockchain via the Internet, locally on their own machine using software such as geth, which syncs a blockchain to a laptop, or through a hosted blockchain (blockchain access provided by a third-party service). Once connected, there are explicit linkages to the main blockchain and multiple test or sandbox blockchain instances for testing smart contracts and other features.

Cryptography, the science of developing data security and privacy algorithms, is utilised on a regular basis to connect new and old blocks. As a result, the blocks that comprise a chain of databases are meant to be immutable and incapable of being modified or altered.

In Bitcoin, a block may contain up to 1 MB of data, suggesting that each block can include thousands of transactions. By definition, blockchain technology is decentralised, which means that the contents of each block, for example, are publicly accessible. Transparency is a key characteristic of blockchain [6].

A blockchain is a decentralised ledger that permanently maintains data across a network of nodes. This not only decentralises information, but also disseminates it. Each node in the network can keep a local copy of the blockchain system, which is regularly updated to ensure consistency across all nodes. A blockchain is a decentralised computing and data sharing platform that enables several nodes to make decisions independently of one another. With centralised systems, single point of failure is a worry. A decentralised system has several coordinate points, which eliminates the single point of failure. Each node collaborates to perform the task in a distributed system. Figure 1 illustrates the fundamental architecture of blockchain. Each user is represented in a distributed network as a node. Each node maintained a copy of the blockchain list, which was regularly updated. A node can perform a range of duties, including transaction initiation, transaction validation, and mining [4].

In a decentralised blockchain network, a node begins a transaction by creating a digital signature using private key cryptography. A transaction is a data structure that encapsulates the exchange of a digital asset between peers on a blockchain network.

Fig. 1 Network architecture
of blockchain technology

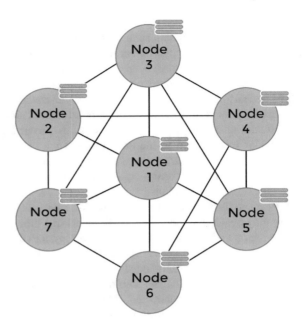

All transactions are held in an unconfirmed transaction pool and distributed over the network using the gossip protocol, a flooding mechanism. Peers are then responsible for selecting and validating these transactions based on predefined criteria. For example, the nodes attempt to validate and authenticate these transactions by determining if the initiator has an enough balance to commence the transaction or by imposing double spending in order to trick the system. When the same amount of money is spent for two or more transactions, this is called double spending. When a transaction has been confirmed and validated by miners, it is put in a block. Miners are peers who pool their computing resources to search for blocks. Miner nodes must solve a computational challenge and consume a large amount of computer resources in order to publish a block. The miner who solves the riddle first wins and gets the opportunity to create a new block. A little payment is given when a new block is successfully formed. The new block is subsequently validated by all of the network's peers using a consensus mechanism, which is a procedure that enables a decentralised network to reach consensus on specific topics. The new block is subsequently included in both the old chain and each peer's local copy of the immutable ledger. At this moment, the transaction is complete. The next block establishes a connection to the freshly generated block through a cryptographic hash pointer. The block is now confirmed for the first time, whilst the transaction is confirmed for the second time. Similarly, each time a new block is added to the chain, the transaction will be reconfirmed. In general, a transaction is deemed complete after six network confirmations. The next part contains a detailed explanation of how the transaction process is carried out [8].

3   Blockchain Transaction Process

A blockchain transaction is a discrete step in the execution of a job that is recorded on the public ledger. These records are sometimes referred to as blocks. These blocks are then created, implemented, and added to the blockchain for verification by all miners on the blockchain network. Any previous transaction's status may be seen at any time, but it cannot be modified.

The blockchain technology at the heart of Bitcoin enables decentralised transactions to take place inside a global peer-to-peer network. As a result, Bitcoin is a decentralised, borderless digital money that is immune to censorship. In general, centralised systems like as banks, in which individuals must place their solemn confidence, may be the primary source of concern. This is where public blockchain technology excels, since it eliminates the requirement for trust between peers when transferring ownership of digital goods. The blockchain is a trustless technology that builds trust by disseminating all network activity. Another consideration before beginning a transaction is security. Security flaws in blockchain mining and consensus methods that significantly rely on a cryptographic hash function can be solved. For example, Bitcoin uses the SHA-256 256-bit secure hash algorithm. Bitcoin can produce 256 bits or 64 characters from any type of input, including text, integers, strings, or even an arbitrarily long computer-generated file. When given the same input, the translated hash result will always be same. On the other hand, a small change in the input completely alters the outcome, which is sometimes referred to as a one-way function, meaning that the input cannot be derived from the output. One can only estimate what the input was, and the probability of guessing right is really high, hence it is safe.

The transaction process begins with establishing the sender's identity, demonstrating that the transaction between the sender and the recipient was initiated by the sender and not by anyone else. The verification procedure is illustrated in Fig. 2 using a simple transaction between A and B as an example. Assume A and B both have a Bitcoin balance and A desires to pay B ten Bitcoins. To transmit the money, A will send a message with the transaction details to the blockchain network. This is accomplished through the use of digital signatures (public and private keys). A provides information about B, including his public address and transaction amount, as well as his public key and digital signature, for the broadcast. A's private key was used to produce that digital signature. Each miner validates transactions separately, based on a variety of characteristics that we shall discuss later in this book. Blockchains make use of the elliptic curve approach for digital signatures (ECDSA). This technique ensures that monies may be used only by their legitimate owners. Each transaction's signature is 256 bits long; hence, to create a fraudulent transaction, a malicious peer/attacker would have to guess 2256 scenarios, which is impossible and wasteful of resources. Not only must the verifier verifies the sender's authenticity, but also the transaction's legitimacy, including whether the sender has sufficient cash to transmit to the receiver. This may be accomplished by consulting the ledger, which has a record of all successful transactions in the past [8].

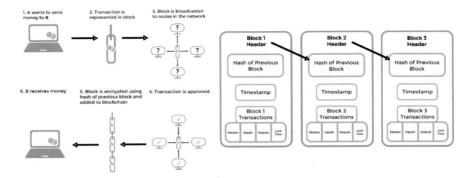

Fig. 2 Transaction between A and B and representation of blocks

Once a block is connected to a blockchain, its contents cannot be modified without impacting all the previous blocks, regardless of the consensus process used [9].

- One of the users initiates a transaction.
- Notifying all blockchain nodes of this request.
- Other nodes analyse the user and the desired transaction.
- If the user and his intended transaction are validated, the new block is added to the chain.
- The transaction has concluded [10].

4   Blockchain Infrastructures

If a blockchain is necessary for a particular application, it must be established whether it is permissionless/public or permissioned/private. A permissionless/public blockchain is one in which anyone can participate (read or write) in it at any time without prior approval, whereas a permissioned blockchain would provide complete transparency into an organisation's internal interactions to anyone with an interest, but not to the general public.

Additionally, certain authority nodes or their counterparts usually permit participation. After deciding on the type of blockchain to use, the framework aids in selecting the form of consensus that is most suited for the application [11].

As seen in Fig. 3, it is grouped into the following categories according on the degree to which participation in blockchain is required:

Blockchain Technology for Public Use: This type of blockchain enables everyone to read and contribute transactions. No prior authorisation is necessary to participate in this type of blockchain. Everyone is invited to participate in the verification process and to join the blockchain network by utilising their own computing capabilities [12].

Without the approval of the other nodes, a node can leave the network. For example, both Bitcoin and Ethereum are public blockchains [13].

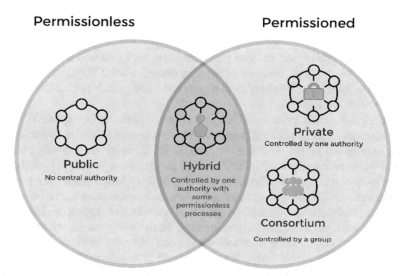

Fig. 3 Blockchain infrastructures

Private Blockchain: This type of blockchain allows just one organisation or all subsidiary organisations within the same group to observe and submit transactions [12]. Private blockchain nodes will be limited; not every node will be allowed to join, and data access will be tightly controlled.

Regardless of the kind of blockchain, its use has a number of advantages. Whilst we may demand public blockchains for convenience, we may also desire private control, such as consortium blockchains or private blockchains, depending on the service we provide or the location where we use it [14].

Community/Consortium Blockchain: In this type of blockchain, many firms form a consortium and are permitted to submit transactions and access transactional data [12].

Consortium blockchains imply that the authoritative node may be selected in advance, and they are frequently utilised in business-to-business transactions. Blockchain data can be public or private, and it is regarded relatively decentralised. Both Hyperledger and R3CEV are consortiums for blockchain technology [14].

A consortium blockchain has two types of users. They are as follows: 1. The users who govern the blockchain and decide who should be permitted access to it. 2. The users of the blockchain. Rather having a single authority, you have several authorities in command here. In essence, you have a group of businesses or individuals cooperating to make decisions that benefit the entire network. These associations are occasionally referred to as consortiums or federations as a result of the term consortium or federated blockchain [15].

Hybrid Blockchain: This is a new category in which transactions can be enabled using any of the three blockchains (public, private, or community/consortium).

Hybrid blockchain enables a blockchain platform to be configured in a variety of ways [12].

A node can participate in both permissionless and permissioned blockchains concurrently to facilitate inter-blockchain communication. This is referred to as a hybrid blockchain. A blockchain platform can handle both permissioned and/or permissionless models [16].

Private blockchains are more suited to corporate applications, especially in regulated industries such as banking, where know-your-customer and anti-money laundering regulations apply. Additionally, private blockchains are generally better at differentiating between private (permissioned) blockchain systems, in which objects have reliable and authorised identities and only strictly vetted parties are permitted to participate, and public (permissionless) blockchain systems, in which objects lack reliable and authorised identities and anyone is permitted to participate [8]. Consortium blockchain systems are frequently used in the commercial sector to record cross-organisational business transactions. In contrast to public blockchain systems, consortium blockchain solutions allow for the participation of only permitted companies in the consensus process. A private blockchain is a network that is distributed but yet centralised and is governed by an organisation or group. Further, subcategories of permissioned blockchain systems include public and private permissioned blockchain systems. In both public and private permissioned blockchain systems, only authorised entities may participate in the consensus process, submit transactions, and maintain the shared ledger. The essential difference is that public permissioned blockchain systems allow anybody to view transactions in the shared ledger, whereas private permissioned blockchain systems restrict access to transactions to approved entities. Permissioned blockchain systems account for the vast majority of blockchain-based commercial applications. Hyperledger Fabric is a representational permissioned blockchain system [17].

As seen in Table 1, numerous techniques to comparing blockchain types are possible.

Determination by Consensus: Whilst all nodes can participate in the consensus process on a public blockchain such as Bitcoin, only a few selected nodes are permitted to confirm a block on a consortium blockchain. The delegates who will determine the confirmed block on the private blockchain will be chosen by a central authority.

Permission to Read: Whilst public blockchain users have read permission, private and consortium blockchains have the ability to restrict access to the distributed ledger. As a result, the organisation or consortium can decide whether or not to make the stored information publicly accessible.

Immutability: Because transactions are recorded in a distributed ledger and validated by all peers in the decentralised blockchain network, editing the public Blockchain is essentially impossible. On the other hand, the consortium and private blockchain ledgers are subject to modification at the discretion of the dominant authority.

Efficiency: The network's scalability is enhanced by the fact that any node on the public blockchain can join or leave the network. However, as the mining

Table 1 Comparison between different blockchain infrastructures [8]

Properties	Category of blockchain		
	Public	Consortium	Private
Nature	Open and decentralised	Controlled and restricted	Controlled and restricted
Participants	Anonymous and resilient	Identified and trusted	Identified and trusted
Consensus procedures	PoW, PoS DPoS	PBFT	PBFT, RAFT
Read/write permission	Permissionless	Permissioned	Permissioned
Immutability	Infeasible to tamper	Could be tampered	Controlled and could be tampered
Efficiency	Low	High	High
Transaction approval frequency	Long (10 min or long)	Short	Short
Energy consumption	High	Low	Low
Transparency	Low	High	High
Observation	Disruptive in terms of disintermediation	Cost effective due to less data redundancy and higher transactions times	Cost effective due to less data redundancy and higher transactions times
Example	Bitcoin, Ethereum, Litecoin, Factom, Blockstream, Dash	Ripple, R3, Hyperledger	Multichain, Blockstack, Bankchain

process grows more sophisticated, and more flexible nodes are added to the network, throughput and latency degrade. On the other hand, private and consortium blockchains can provide improved performance and energy efficiency due to their fewer validators and customizable consensus methods.

Centralisation: Whilst the public blockchain is decentralised, the consortium blockchain is somewhat centralised, and the private blockchain is overseen by a centralised body. Due to the fact that public blockchain is available to everyone on the planet, it has the potential to attract a sizable user base. Additionally, communities are quite active. Daily, new public blockchains are created. The consortium blockchain has a number of potential commercial uses. Hyperledger is currently creating blockchain frameworks for corporate consortiums. Additionally, Ethereum has made tools for establishing consortium blockchains available. Numerous firms continue to utilise private blockchains for increased efficiency and auditability [8].

5   Consensus Algorithms

The Byzantine general's (BG) problem, which was raised in, is a challenge about how to achieve consensus in a blockchain amongst untrustworthy nodes. In the BG problem, a group of generals commanding a division of the Byzantine army circle the city. Certain generals would want to attack, whilst others would prefer to retire. If, on the other hand, only a part of the generals attack the city, the attack will fail. As a result, they must choose between assaulting and retreating [18]. This is seen in Fig. 4. Consensus is difficult to obtain in a scattered context. This is also a concern for blockchain technology, as the network is distributed. There is no central node in blockchain that ensures the identicality of all distributed node ledgers. Many protocols are necessary to ensure the consistency of ledgers across multiple nodes. Following that, we will discuss several potential methods for achieving blockchain consensus [19].

5.1   Approaches to Consensus

5.1.1   Proof of Work (POW)

The Bitcoin network makes use of a consensus process called proof of work (PoW). In a decentralised network, someone must be appointed to record transactions. The simplest way is random selection. On the other side, random selection is vulnerable to attack. As a result, before a node may publish a block of transactions, it must prove to the network that it is unlikely to be attacked. According to the Bitcoin whitepaper,

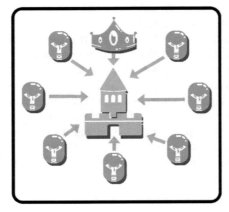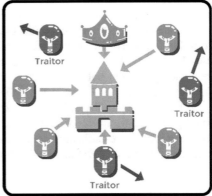

Coordinated Attack Leading to Victory   Uncoordinated Attack Leading to Defeat

Fig. 4   Coordinated and uncoordinated attack leading to defeat

the PoW system works by looking for values that have a hash starting with a number of zero bits. This is accomplished by attaching a nonce (a piece of code that does work) to the original value until the resultant hash has the appropriate number of zero bits. Once this nonce is chosen and the proof of work is satisfied, the block cannot be updated without redoing all of the previous work for that and subsequent blocks.

5.1.2 Proof of Stake (PoS)

Original Proof of Stake

Proof of stake (PoS) is a hybrid architecture that employs proof of work (PoW) for initial currency minting and proof of stake (PoS) for the bulk of network security. It was originally used in 2012 as a cryptocurrency called Peercoin. In a PoS system, the age of each currency is measured in "coin-days." This concept is readily shown with the following example: ten coins stored for ten days equals one hundred coin-days. Each time one of these currencies is used in a transaction, its age is decreased and reset to zero. Unlike in a PoW system, where the chain that performs the most work is called the main chain, in a PoS system, the chain that consumes the most coins is considered the main chain [20].

Under the PoS technique, a validator pays himself (and so spends his coin age) for the right of minting a new block for the network. Under the following situations, the system determines the minimum amount that a validator must contribute in order to mint a new block:

proofhash coins target age.

According to the whitepaper of another proof of stake cryptocurrency called BlackCoin, the proofhash is the obfuscation sum that is based on a stake, unspent output, and the current time. Coins represent the number of coins a miner has paid for mining rights, age represents the age of those coins, and aim represents the amount of coins required by the network via a network, a difficulty adjustment approach similar to that used in PoW. The difficulty of the challenge is due to the intricacy of the mathematical problem that validators must answer in the case of blockchains.

The proof of stake (PoS) protocol is a more energy-efficient variant of the PoW protocol. Miners must establish that they own the money in a proof of stake (PoS) system. Individuals having a greater number of currencies are believed to be less likely to attack the network. Because the wealthiest people are almost guaranteed to be prominent in the network, selecting individuals only on the basis of their account balance is very unfair. As a result, a number of approaches are offered for selecting the next block to forge utilising a mixture of stake sizes. Blackcoin, in particular, forecasts the future generation using randomisation. It implements a formula that takes into account both the lowest hash value and the stake size.

Peercoin likes to choose currencies on the basis of their age. In Peercoin, older and larger groupings of coins have a greater probability of mining the next block. PoS is more energy efficient and saves more energy than PoW. Unfortunately, because mining is so inexpensive, attacks may occur. Numerous blockchains begin as PoW and gradually migrate to PoS. For example, Ethereum is considering switching from Ethash (a form of PoW) to Casper (a type of PoS) [18].

Delegated Proof of Stake (dPOS)
Stakeholders choose to elect any number of witnesses to create blocks in a delegated proof of stake (dPOS) system. The roster of witnesses is shuffled during each maintenance period, and each witness is given a chance to produce a block at a set rate of one block every n seconds (where n depends on the implementation). Witnesses are compensated for each block produced; but, if they fail to produce a block after being elected, they may be voted out in subsequent elections.

Blocks are generated every three seconds by approved producers on the EOS blockchain, and the list of those producers is rotated every 21 blocks. If a producer has not created a block in the last 24 h, they are removed from consideration until they inform the blockchain of their desire to resume block production [21].

5.1.3 Practical Byzantine Fault Tolerance

In the late 1990s, Barbara Liskov and Miguel Castro [22] devised the practical Byzantine fault tolerance consensus approach. pBFT was designed to work effectively in asynchronous systems (where there is no time constraint on obtaining a response to a request). It has been designed to use the least amount of overhead time possible. Its objective was to address a number of concerns with currently available Byzantine fault tolerance methods. Two application cases are distributed computing and blockchain.

Byzantine fault tolerance (BFT) is a property of distributed networks that enables them to achieve consensus (agreement on the same value) even when some nodes in the network fail to respond or respond with incorrect information. A BFT mechanism's objective is to defend against system failures by leveraging collective decision-making (both correct and incorrect nodes) with the purpose of minimising the influence of defective nodes. BFT is derived from the Byzantine generals' problem.

Practical byzantine fault tolerance (PBFT) is a method for replicating byzantine fault tolerance. Hyperledger fabric employs PBFT as its consensus technique since it can resist up to 1/3 malicious byzantine copies. Each round determines a new block. Each round, a primary would be picked according to certain parameters [23]. Additionally, it is responsible for the transaction's ordering. The technique is divided into three phases: preparation, preparation, and commitment. Each phase, a node progresses if it obtains votes from more than two-thirds of all nodes. As a result, PBFT requires that the network understands each node. SCP is a Byzantine agreement mechanism akin to PBFT. In PBFT, each node is forced to challenge other nodes, but in SCP, participants can choose which set of other participants to trust. Antshares' dBFT is based on the delegated Byzantine fault tolerance (PBFT). A subset of the dBFT network's professional nodes is chosen to record transactions.

5.1.4   Tendermint

Tendermint is a method for achieving Byzantine consensus. Each round determines a new block. A proposer would be picked in this round to broadcast an unconfirmed block. It may be summarised in three steps: The first step is to prevote. Validators choose whether or not to air a prevote on the proposed block. The second step is to make a pre-commitment. If the proposed block receives more than two-thirds of the prevotes, the node broadcasts a precommit for it. When the node has received more than two-thirds of the precommits, the commit procedure begins. (3) Carry out the process. The node validates the block and then broadcasts a commit for it. If it has received two-thirds of the commits, the node accepts the block. In contrast to PBFT, nodes must lock their currency before becoming validators [8]. If a validator is determined to be dishonest, he or she will be punished.

6   Characteristics of Blockchain Technology

Numerous aspects characterise blockchain systems. The next sections discuss critical characteristics in terms of distributions, implementation, and functionality.

The Centralisation and Decentralisation of Government: In most centralised database systems, transactions are intrinsically trusted or approved by central trusted intermediaries that ensure their legitimacy. When central servers are employed, more costs are incurred, and performance becomes a critical factor to consider. Blockchain technology is a viable solution for distributed transaction management challenges in a peer-to-peer network. The extent to which an owner retains control over node selection determines the degree of decentralisation. This setup is comparable to a typical distributed database system. The status of nodes as members is regulated by consortium policies. All nodes preserve blockchain information, and no single party has total control over the system. Decentralisation enhances all forms of blockchains to varying degrees, by reducing the single point of failure and ensuring data integrity.

Persistency: Persistent transactions are those that are kept in a blockchain ledger because they are dispersed over the network, with each node preserving and controlling its own records. Persistence is only possible if the majority of nodes are benign. Transparency and impermanence are two characteristics derived from this feature (temper resistance). Due to their openness and immutability, blockchains are auditable [24].

Validity: Unlike certain distributed systems, blockchains do not need execution from each node. In a blockchain system, transactions, or blocks, are broadcast and confirmed by other nodes. As a result, any forgery would be immediately detected [24].

Anonymity and Identity: One of the primary characteristics of public blockchains is their anonymity. In this approach, a user's identify is unconnected to their real-world identity. A person may create many identities in order to protect themselves from identity theft. Personal information does not require central monitoring. As a consequence, real-world identification cannot be determined merely from transaction data, so preserving some level of anonymity. By contrast, identification is often necessary in systems that are maintained and managed by identifiable entities, such as private and permissioned blockchains.

Auditability: Through the use of a timestamp and permanent information, nodes in a blockchain network may easily verify and track records. The degree of auditability varies according to the type of blockchain system and how it is deployed.

Closed and Open Blockchains: Open blockchains rely on public nodes to maintain transaction records. Closed blockchains rely on private nodes to maintain the ledger. As a consequence, anybody may publish a transaction and join the system by following a set of rules, and the data held inside this blockchain is accessible to the whole globe. Permissioned blockchains are semi-open in nature due to the requirement for nodes to be pre-specified or verified prior to joining. They exist in the grey area between public and private blockchains. The information included in this blockchain is subject to the consortium's regulations, which may dictate whether the information is totally open, partially open, or closed. As with permissioned blockchains, private blockchains are governed by laws that govern how nodes are picked and the degree of data openness. However, they are dependent on a single firm or owner.

7 Advantages and Challenges of Blockchain Technology

Blockchain technology, one of the most significant advancements in recent years, will affect a wide variety of industries in the next years. Centralised technologies provide significant difficulties in our daily lives. Blockchain technology, which enables decentralised organisations, apps, and transactions, has ushered in a revolution as a solution to these difficulties. It enables capabilities that are not available in a centralised system, such as anonymous users who are not required to provide their identities and the ability for anybody to view the processes. Additionally to these benefits, there are also drawbacks, including the possibility of becoming a centre again in the future, expensive operation expenses, and transaction speeds. All of them are discussed in further detail in the following subheadings.

7.1 Challenges of Blockchain Technology

Transparency—Because of blockchain, transaction histories are more transparent than they have ever been. Because this is a distributed ledger, each node in the network has a copy of the documentation. A blockchain ledger's data is public and may be read by anybody. If a transaction's history changes, everyone on the network will see the change and the updated record.

Security—By and large, blockchain excels all other methods of record keeping in terms of security. Consensus is required to update and/or alter a blockchain network's shared transaction documentation. The information is updated only when all or a majority of nodes agree. Additionally, after a transaction is accepted, it is encrypted and connected to the transaction that preceded it. As a result, no one individual or political party has the authority to alter a record. Due to the decentralised nature of blockchain, no one has the authority to edit records at their leisure. Blockchain technology may be used to apply rigorous security measures in any sector that requires the protection of sensitive data, such as government, healthcare, and financial institutions.

Efficiency—Human error is a frequent occurrence in paper-based operations. Simultaneously, conducting any operation using this approach results in a loss of time and money. These operations can be expedited using blockchain technology, and the process can be made more efficient by reducing the possibility of error. Due to the fact that the blockchain system uses a single immutable ledger that can be read by everyone, everyone has access to the same information without the need for parties to maintain multiple ledgers.

Traceability—Tracing items back to their origins can be challenging in complicated supply networks. However, with blockchain, all commodities transactions are recorded, providing an audit trail for determining the origin of a certain item. Additionally, you will learn about each stop the products make along the road, and this degree of product tracking may assist in establishing authenticity and avoiding fraud.

Cost savings—Because blockchain technology eliminates the need for third parties and middlemen, it results in significant cost savings. Cost and time savings are realised when updates are made, as all users have access to the same, unmodified version of the ledger at all times.

7.2 Disadvantages of Blockchain Technology

Despite its great promise, blockchain technology still faces substantial barriers to mainstream adoption. The following are a few of the difficulties experienced.

Scalability—Due to the rising volume of transactions on a daily basis, the size of blockchain systems is regularly expanding. Each node in the blockchain network must store all blockchain data in order to perform verification and consensus procedures. As a result, the blockchain continues to grow in complexity. The number of transactions verified on the chain in a given period of time is restricted due to constraints such as block capacity and block publication speed. For example, the Bitcoin blockchain can confirm around seven transactions per second. Bitcoin is extremely slow in comparison with other payment systems. For example, the VISA network (VisaNet) has a transaction processing capability of 65,000 per second. Apart from the limits imposed by the blockchain's block structure, there may be delays in transaction speeds as a result of miners prioritising big money transfers in order to earn a high profit and disregarding minor transactions. The fact that each node in the network maintains and continually updates all blockchain data complicates the system's operation [8].

Storage—A storage issue arises as a result of blockchain databases being retained eternally on all network nodes. The database will expand in size as the number of transactions rises, and home computers will never be able to manage an infinite amount of data that is simply added. For reference, the Ethereum blockchain is expanding at a 55 GB per year rate [25].

Privacy—Users can transact on the blockchain using their own public and private keys without disclosing their genuine identities, and user privacy can be protected to a significant extent. Simultaneously, because transaction transparency is crucial on the blockchain, information about the sender, receiver, time stamp, and transferred value is publicly available. This enables consumers to access all transaction records, as well as information about the merchants with whom they shop and their account balance. It is feasible to determine a user's genuine identify through third-party inspection of this shared data [8].

Regulations—Technical/scalability difficulties, business model challenges, scandals and public perception, government laws, and personal data privacy concerns are all possible roadblocks to blockchain's widespread and deeper implementation. Before blockchain to become a reality in the financial services sector, it must overcome 10 significant hurdles. Cost–benefit analysis, cost mutualisation, incentive alignment, standardisation, scalability, governance, legal issues, security, simplicity, and regulatory interventions are a few of these. Legislation and regulation may have an effect on how far and how quickly technology advances [8].

Security—Due to the immaturity of technology, there exist flaws that expose people to cybercrime. 51% attacks are one of the most well-known blockchain security flaws. In a 51% assault, one or more malicious actors seize majority control of a blockchain's hash rate. If they hold the majority hash rate, they can reverse transactions to conduct double-spends and prevent other miners from confirming blocks [8].

Selfish mining is another uneven method of boosting block rewards in mining pools that jeopardises a blockchain network's integrity. Despite the fact that hostile nodes with over 51% computing power may gain control of the blockchain network, showed a blockchain network that is still vulnerable if someone with a little amount of hashing power chooses to cheat. Selfish mining can be initiated by a single miner or a group of miners who do not broadcast verified blocks to the rest of the network. Then, to maintain their advantage, they continue mining for the next block. The solution blocks are made available only once certain conditions have been satisfied. As a consequence, the selfish miner's chain lengthens and becomes more sophisticated, forcing the network to embrace their solutions whilst other miners spend resources on a pointless branch. Finally, the greedy miners demand further funds. This incentivizes rational miners to join the longer chain, potentially boosting the selfish pool's strength to beyond 51%.

Numerous well-known blockchain platforms have proved their resilience to assaults and their lack of significant faults. However, the applications built on top of them (for example, smart contracts) remain subject to security flaws that might have disastrous consequences. Prospective users will remain hesitant until these security concerns are resolved, hence delaying widespread adoption.

Keys to the kingdom- Blockchain technology utilises public-key (or asymmetric) cryptography to ensure that users retain control of their bitcoin units (or any other blockchain data). Each blockchain address is connected with a private key. Whilst the address can be shared, the private key should be kept secret. Users may access their funds solely through their private key, thereby turning them into their own bank. If a user loses their private key, they effectively lose their money and have no recourse [25].

8   How to Determine Whether You Need Blockchain Technology

There are hundreds of startups that have started to develop various business solutions using blockchain technology, as well as consortia formed by various large banks and companies aiming to develop solutions in the field of blockchain. There are also opinions that stay a little more distant from these developments and question the high expectations formed around blockchain technology and suggest focussing on projects where the net benefits of blockchain technology can be seen. Blockchain technology is not suitable for every use case and business model. Therefore, before using the blockchain, it should be checked whether it complies with the requirements. For example, there is no need to use blockchain if the data to be kept will not be created and read by more than one person. However, since the blockchain provides the distribution of trust between the parties, there is no need for a trust mechanism between the parties that will process/use the data, even if there is no need for the blockchain. Blockchain technology will not be needed if there is no need for a public auditing mechanism and changing records is not critical [25].

If blockchain technology is desired to be used after the above decisive criteria, the following benefits can be provided in the system to be used:

Since the trust mechanism is not based on individuals or institutions, but on multiple nodes in a distributed structure and the difficulty of the underlying mathematical operations, technical trust is provided. Depending on the structure of the blockchain, the privacy of the users can be fully ensured in the system to be established. By eliminating the need for a centralised system, a system that manages itself is obtained. The system can be created and maintained not by one or a few authorities, but by many small users.

9 Blockchain Technology Applications in Industries

Due to the potential benefits for a wide variety of organisations, blockchain technologies have recently moved to the top of the scientific and industrial agendas. This is due to their practical abilities in overcoming a large number of the issues inhibiting advancement in a wide variety of industrial industries. Amongst these issues are the safe collection and exchange of transactional data, the automation and efficiency of supply chain activities, and the enhancement of transparency throughout the value chain. Blockchain technology addresses these challenges well by utilising distributed, shared, secure, and permissioned transactional ledgers [26]. The utilisation of blockchain technology and their adaptability to a number of situations enables a broad variety of industrial applications, including increased efficiency and security, enhanced traceability and transparency, and cheaper pricing. This article discusses many industrial applications for blockchain technology. This article discusses the benefits, drawbacks, and limits of implementing blockchain technology in a variety of industrial applications. Additionally, the essay seeks to explain the conditions for blockchain implementation in a variety of industrial applications. The research reveals that there are numerous opportunities for blockchain implementation in a variety of industrial sectors; nevertheless, some barriers must be addressed in order to expand the technology's adoption.

We investigate the usage of blockchain technology in a variety of industries. This section discusses the industrial applications of blockchain technology, as well as its advantages and disadvantages. The applications covered are grouped by the several industrial sectors they serve, including banking, healthcare, logistics, manufacturing, energy, agriculture and food, robots, and entertainment. Additionally, the essay examines the primary conditions for implementing blockchain in industrial applications, as well as some of the sector's outstanding issues.

Blockchain technology has emerged as a promising technology as a result of the introduction and widespread use of cryptocurrencies. However, it is shown indications of maturing into a viable technology with a broad range of applications in a variety of disciplines. This section discusses multiple industrial applications of blockchain technology, all of which provide several commercial and industrial benefits. These applications are classified according to their subject matter. Figure 5 summarises the primary benefits of using blockchain in the various industrial sectors mentioned.

9.1 Financial Industry

Because of blockchain's success in supporting cryptocurrencies, it was only natural for the financial sector to follow suit with blockchain applications in other financial fields. In general, trusted third parties are employed to perform financial transactions

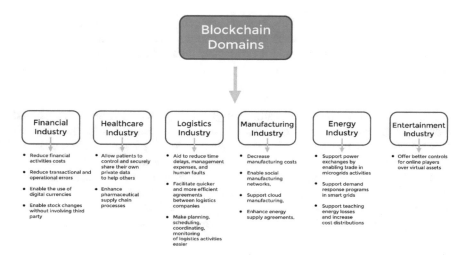

Fig. 5 Blockchain and its applications in various industries

between individuals and businesses [27]. These third parties do four tasks: 1. Verifying the validity of deals 2. Avoiding duplicate financial transactions 3. Keeping track of and confirming financial transactions 4. Providing help to clients or associates as agents two of these functions can be replaced by blockchain: preventing transaction duplication and registering and verifying financial transactions. It is simple to restrict, for example, a client from making numerous payments totalling more than what they owe using blockchain. In fact, normal cheques might be used to execute this conduct unlawfully. However, this is hard to execute with blockchain since all financial transactions must be confirmed collectively before being carried out [26].

Simultaneously, blockchain can serve as a secure register for completed financial transactions. After being added to the chain, this registration cannot be changed by any of the parties involved. It may also be used to verify and confirm the transactions that have been completed through group checks and verifications. Many financial applications, such as the ones below, are made possible by these two properties.

9.2 Healthcare Industry

Patient data is a very sensitive and critical element of healthcare. Frequently, a patient's medical records are spread amongst many systems owned and maintained by one or more healthcare providers. The digital revolution introduced the possibility of converting patient data into what is often referred to as an electronic medical record (EMR). Exchange of EMRs across multiple healthcare providers and healthcare-related organisations is challenging due to a variety of issues, including security and privacy. The blockchain technology may be used to ease the safe exchange of electronic medical records (EMRs) and other forms of healthcare data between several providers. Gem, a blockchain-based startup, has built a network for developing healthcare applications and a centralised data infrastructure for healthcare. Additionally, Tierion, a startup company, developed a platform for the storage of healthcare data. Additionally, this technology facilitates the verification and auditing of healthcare data and processes [26].

Another research offers the health data gateway (HDG) as a blockchain-based application architecture that would enable patients to securely share and manage their health data whilst maintaining their privacy. Whilst patients retain ownership and management of their health data, they are unable to modify, delete, or add any health-related information. In this design, access rules may be set using software services, and other organisations can access the EMRs via other services. Due to the fact that EMRs may be amended by a variety of authorised entities, blockchain technology can be utilised to track all of these modifications and offer a trustworthy audit trail that includes step-by-step documentation of all EMR updates [27].

9.3 Logistics Industry

Logistics management apps are a type of software that aids in the management of raw material, product, and service deliveries between producers/sellers and consumer destinations. These may all be associated with the same organisation or may be scattered across several organisations and entities. These applications can benefit significantly from the blockchain's amazing support. One of the logistical management issues is including several organisations in the activities. This may also include a range of coordinated sub-activities conducted by other businesses, including as manufacturers, storage companies, shipping companies, and regulatory bodies. Any logistics management programme must have the ability to plan, schedule, coordinate, monitor, and validate activities.

Such functions may be supported by blockchain technology in an efficient and safe manner. Time delays, administrative costs, and human error may all be minimised by verifying, recording, and auditing logistics transactions utilising blockchain's shared distributed ledgers. Additionally, utilising smart contracts would enable firms to negotiate agreements and build legally binding contracts more quickly and at a lower cost. Due to these benefits, blockchain is likely to have a major impact on the logistics industry. Numerous firms are entering this market with blockchain-based logistics management systems and solutions. Provenance is one type of traceability system that links consumers and suppliers for a variety of logistical tasks. Hijro is another example, as it provides a platform for global supply chain management applications.

9.4 Manufacturing Industry

The manufacturing industry is making major strides towards smart manufacturing and automated/autonomous processes, and blockchain technology has the potential to assist in a variety of ways. Logistics management, which we previously discussed, is a critical field. Logistics management is crucial for any industry that wishes to assure fair pricing and timely delivery of raw materials and supplies used in manufacturing. Additionally, it aides in the efficient and timely delivery of their products in order to satisfy their consumers' requests. As with any other logistics management system, utilising blockchain for manufacturing logistics management may help reduce time delays, management costs, and human errors. As a consequence, organisations may be able to reduce production costs whilst increasing their agility and competitiveness [28].

Blockchain technology might be utilised to assist in cloud manufacturing. Cloud manufacturing is a new manufacturing paradigm that leverages cloud computing, the Internet of things (IoT), service-oriented computing, and virtualization to turn manufacturing resources and processes into a network of intelligently linked and managed manufacturing services. In this context, it is recommended that blockchain technology be used to create a secure decentralised cloud manufacturing architecture and to facilitate secure information sharing for manufacturing design, such as injection moulding design and redesign.

Genesis of things is a startup company developing a platform that integrates 3D printing, blockchain technology, and the Internet of things to enable more creative manufacturing processes. This enables more efficient customisation by lowering the cost of custom 3D printed objects and automating several manufacturing and related operations [27]. Another advantage is the ability for cooperating manufacturers to safely exchange and share manufacturing data. In these cases, data must not only be safeguarded, but also tracked and regulated to ensure appropriate access and avoid illegal changes or tampering [26].

9.5 Energy Industry

Microgrids are one of the most prevalent applications of blockchain in the energy sector. A microgrid is a locally linked and managed network of electric power sources and loads with the objective of increasing the efficiency and reliability of energy production and consumption [26]. Electricity can be generated via distributed power generators, renewable energy stations, and energy storage components installed in facilities developed and maintained by a variety of companies or energy suppliers. One of the primary advantages of microgrid technology is that it enables home-owners and other electric power consumers, such as businesses, to not only obtain the energy they require, but also to generate and sell surplus energy to the grid. Blockchain technology may be used to streamline, record, and confirm power selling and purchasing transactions in microgrids. For example, blockchain technology is being used to power a microgrid of 130 buildings in Brooklyn, New York. This significantly decreases or eliminates the requirement for intermediaries to facilitate energy sales and purchases between these buildings. Blockchain technology may also be utilised to measure energy losses caused by energy transfers in islanded microgrids [27]. As a result, the real location of power and its losses across the grid network, as well as the costs assigned to participants, are more precisely linked.

References

1. Fernandez-Carames TM, Fraga-Lamas P (2019) A review on the application of blockchain to the next generation of cybersecure industry 4.0 smart factories. Ieee Access 7:45201–45218
2. Treiblmaier H (2020) Toward more rigorous blockchain research: recommendations for writing blockchain case studies. In: Blockchain and distributed ledger technology use cases. Springer, Cham, pp 1–31
3. Paulavičius R, Grigaitis S, Igumenov A, Filatovas E (2019) A decade of blockchain: review of the current status, challenges, and future directions. Informatica 30(4):729–748
4. Mohanta BK, Jena D, Panda SS, Sobhanayak S (2019) Blockchain technology: a survey on applications and security privacy challenges. Internet of Things 8:100107
5. Mueller P (2018) Application of blockchain technology. It-Inf Technol 60(5–6):249–251
6. Ashurst S, Tempesta S, Kampakis S (2021) Blockchain applied: practical technology and use cases of enterprise blockchain for the real world. Productivity Press
7. Barenji RV (2021) A blockchain technology based trust system for cloud manufacturing. J Intell Manuf 1–15
8. Monrat AA, Schelén O, Andersson K (2019) A survey of blockchain from the perspectives of applications, challenges, and opportunities. IEEE Access 7:117134–117151
9. Oliva GA, Hassan AE, Jiang ZMJ (2020) An exploratory study of smart contracts in the Ethereum blockchain platform. Empir Softw Eng 25(3):1864–1904
10. Hasankhani A, Hakimi SM, Bisheh-Niasar M, Shafie-khah M, Asadolahi H (2021) Blockchain technology in the future smart grids: a comprehensive review and frameworks. Int J Electr Power Energy Syst 129:106811
11. Gourisetti SNG, Mylrea M, Patangia H (2019) Evaluation and demonstration of blockchain applicability framework. IEEE Trans Eng Manage 67(4):1142–1156
12. Shrivas MK, Yeboah DT (2018) The disruptive blockchain: types, platforms and applications. In: 5th Texila World Conference for Scholars (TWCS) on transformation: the creative potential of interdisciplinary
13. Maharjan PS (2018) Performance analysis of blockchain platforms doctoral dissertation. University of Nevada, Las Vegas.
14. Niranjanamurthy M, Nithya BN, Jagannatha S (2019) Analysis of Blockchain technology: pros, cons and SWOT. Clust Comput 22(6):14743–14757
15. Hughes A, Park A, Kietzmann J, Archer-Brown C (2019) Beyond Bitcoin: what blockchain and distributed ledger technologies mean for firms. Bus Horiz 62(3):273–281
16. Rennock MJ, Cohn A, Butcher JR (2018) Blockchain technology and regulatory investigations. Pract Law Litigation 35–44
17. Xie J, Tang H, Huang T, Yu FR, Xie R, Liu J, Liu Y (2019) A survey of blockchain technology applied to smart cities: research issues and challenges. IEEE Commun Surv Tutor 21(3):2794–2830
18. Zheng Z, Xie S, Dai HN, Chen X, Wang H (2018) Blockchain challenges and opportunities: a survey. Int J Web Grid Serv 14(4):352–375

19. Barenji RV, Nejad MG (2022) Blockchain applications in UAV-towards aviation 4.0. In: Intelligent and fuzzy techniques in aviation 4.0. Springer, Cham, pp 411–430
20. Lai R, Chuen DLK (2018) Blockchain–from public to private. In: Handbook of blockchain, digital finance, and inclusion, vol 2. Academic Press, pp 145–177
21. Namasudra S, Deka GC (eds) (2021) Applications of blockchain in healthcare. Springer, Singapore
22. Castro M, Liskov B (1999) Practical byzantine fault tolerance. In: OSDI, vol 99, no 1999, pp 173–186
23. Vatankhah Barenji R (2021) A blockchain technology based trust system for cloud manufacturing. J Intell Manuf 1–15.
24. Viriyasitavat W, Hoonsopon D (2019) Blockchain characteristics and consensus in modern business processes. J Ind Inf Integr 13:32–39
25. Maruti Techlabs (2021) Blockchain—benefits, drawbacks and everything you need to know (online). Available from: https://marutitech.com/benefits-of-blockchain/
26. Al-Jaroodi J, Mohamed N (2019) Blockchain in industries: a survey. IEEE Access 7:36500–36515
27. Saadati Z, Zeki CP, Vatankhah Barenji R (2021) On the development of blockchain-based learning management system as a metacognitive tool to support self-regulation learning in online higher education. Interact Learn Environ 1–24
28. Porter ME, Heppelmann JE (2019) HBR's 10 must reads on AI, analytics, and the new machine age

Blockchain Technology Application in Manufacturing

Melis Etim and Egemen Akçay

1 Introduction

As the name implies, a blockchain is a series of data blocks. It is a distributed database that allows for the tracking of data that has been encrypted using a particular technique called hash. Although blockchain was initially developed for bitcoin trading, its promise extends well beyond cryptocurrencies.

Numerous structural changes have occurred as a result of the development and widespread use of technology in the modern era. Peer-to-peer communication enables the establishment of a decentralised environment in the pioneering of the Internet network and software architecture.

In this context, we may explore a novel technique that utilises a distributed system structure and encrypted data to enable peer-to-peer communication without the use of a central server. If time and money are wasted as a result of the tools employed in the process, we may argue that blockchain technology can be used. Blockchain technology may be used to a variety of various areas and business lines, including supply chain management, energy, and healthcare.

The purpose of this research is to teach some fundamental concepts about blockchain technology and its applications.

The block chain's application examples are discussed, as well as the missing features of the blockchain from a variety of perspectives and potential future concerns.

M. Etim (✉) · E. Akçay
Department of Industrial Engineering, Hacettepe University, Beytepe Campus, 06800 Ankara, Turkey
e-mail: melisetim@gmail.com

© The Author(s), under exclusive license to Springer Nature Singapore Pte Ltd. 2023   117
A. Azizi and R. V. Barenji (eds.), *Industry 4.0*, Emerging Trends in Mechatronics, https://doi.org/10.1007/978-981-19-2012-7_5

2 Application of Blockchain

2.1 Smart Contract

Nowadays, the majority of businesses are connected with partners and suppliers across numerous industries. The common currency for all of these things is data, but data in real-world infrastructure is at danger of being compromised by errors, malicious actors, and conflict.

Smart contracts are a powerful response to the real-world active management of data in a safe, access-controlled, and open manner. Smart contracts enable contemporary, integrated organisations, their partners, suppliers, and regulators to collaborate without compromising their integrity, since smart contracts automatically execute business rules when certain circumstances are satisfied.

A smart contract's distinguishing characteristic is its capacity to reduce complex business rules between parties to a manageable set of automated or semi-automated processes. Because each smart contract is cryptographically secure, the results of the executed rules are also rendered as highly secure, immutable data that cannot be changed. Thus, the smart contract's net impact is to foster trust: once all parties have agreed to the business rules described in the smart contract, they may have confidence that the automation of the smart contract and the blockchain will carry out the agreed-upon results.

2.2 Private Network

The decentralisation specialty of the blockchain network is very useful for the sectors that are needed security prevention such as medical, military, and economy. Privacy comes with the nature of the blockchain network. Basically, private blockchain networks are permissioned blockchains and works on basis of access controls who will participate in the network.

For example, if a model that predicts the likelihood of re-admission for a certain group of patients is constructed using data from multiple institutions, it will be more generalizable. Whilst there are privacy-protecting approaches for building predictive models, most models are built on a centralised architecture, which introduces security and robustness risks such as single-point-of-failure or malicious data change. This can be a good adaptation chance from centralised architecture to blockchain technology for privacy-preserving machine learning [1] (Fig. 1).

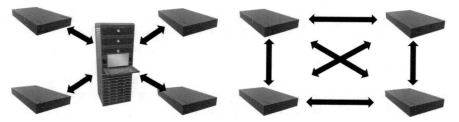

Fig. 1 Centralised and decentralised architecture

3 Case Studies

There are many application in the sector 2 of blockchain applications are being examined in detail further topics.

3.1 Trust System for Cloud Manufacturing

As an application, Reza Vatankah's "blockchain technology-based trust system for cloud manufacturing" article will be examined in this topic.

3.1.1 Cloud Manufacturing

Increased market competitiveness (particularly in low-wage countries) motivates manufacturing businesses to research and embrace innovative production and business models, as well as to change their long-term supply chains to perhaps shorter-term larger manufacturing networks [2]. The trend is towards sectors in which capabilities may be accessed and used from any location on Earth at any time [3]. Recently, CM has been presented [4–6]. It was created to amass and combine vast, valued skills [7] from several disciplines and industrial enterprises for the benefit of customers. In this paradigm, capacities are mainly independent, scattered geographically, and diverse in terms of their working environment, culture, social capital, and aspirations [8]. It is a concept that enables the manufacturing firms to connect to a common pool of flexible production capabilities via an omnipresent, ubiquitous, beneficial, and on-demand network connection.

ICT infrastructure plays a critical enabling role in the CM model by providing information exchange/sharing, secure communication, coordination, and collaboration services. It acts as the hero of a "operating system" or executor, concealing the identities of the organisations involved [9]. It maintains distributed services with many owners and adheres to service-oriented architecture (SOA) principles by utilising three primary entities: "service provider," "service manager," and "service requester" [10, 11]. It improves information flow, increases the flexibility of corporate operations, and enables collaboration amongst diverse parties. "Service manager" includes a capability pool (CP) that aggregates virtualized capabilities from a variety of stakeholders regardless of their enterprise resource planning (ERP) or manufacturing execution system (MES) kind or brand [12]. CP is an environment that hosts virtualized capabilities given by many stakeholders and makes them available to other businesses (i.e., customers). It establishes a link between requesters (individuals or businesses) and suppliers (e.g., manufacturing enterprises, supplies).

3.1.2 Blockchain Technology

BCT is a decentralised open ledger technology that enables the recording and sharing of information by a community [13, 14]. Each participant retains his or her own copy of the information, and any revisions must be validated collaboratively. The data might represent a transaction, a contract, an asset, an identity, or almost anything else that can be defined digitally. Because BCT entries are permanent, transparent, and searchable, community members may access whole transaction histories. Each collection of transactions constitutes a new "block" at the conclusion of a "chain." A protocol governs the process of initiating, validating, recording, and disseminating new adjustments or additions. To connect the records linked in the blocks, cryptographic validation is utilised.

3.1.3 State of the Art

Trust in Collaboration

Historically, trust has developed in business partnerships in one or more of the following ways: (A) It can occur between local organisations through the examination of one another before to, during, and after engagement. (B) Through connections and regular encounters, multinational corporations may develop trust. For example, an organisation in position A (e.g. AA) works with another entity in location B (e.g. BB), AA and BB regularly communicate with one other and grow to trust each other. (C) Trust is established between two entities through the use of local peers. For instance, because AA and BB entities trust one another, additional entities in positions A and B (for example, XA and XB) gain knowledge about AA and BB through their interactions. (D) Trust may be lent to other entities. For instance, organisations are likely to be a joint venture with another well-known organisation (E), and trust can be built over many years through the creation of brand names.

The article's central premise is that trust may be established as a result of the feedback value distributed, saved, and assigned to each provider and requester in the network via BCT. Several strategies were used to uncover certain insurmountable challenges for the CM platform:

- As a fully computerised system, CM cannot make the assumption that reputation is inherently reliable.
- This poses several challenges, including scalability, exercising CM's centralised authority, maintenance overhead, managing denial of service attacks, and fraud.
- To address the aforementioned issues, this article presents a trust-based approach for the CM prop of BCT.

Proposed Cloud Manufacturing Framework Embedded with Blockchain Technology

The embedded CM architecture and block trust are depicted in Fig. 2. The framework is composed on three major components: the digital certificate issuance unit (DCIU), the talent pool unit (CaPU), and the service-oriented architecture (SOA), all of which are powered by the digital firm unit (DFU). The DFU acts as both a service requestor (consumer) and a service provider, whereas the DCIU acts as a service provider and the CaPU acts as a service manager. A new firm (i.e., user) must first access and register in the system. During the registration process, the CAPU must receive information on the firm's technical capabilities, general information, and the previous interactions. Any company that owns a DC qualifies as a DFU for the purposes of requesting/providing services to/from the CaPU. It communicates with manufacturing execution systems (MESs) and enterprise resource planning (ERP) systems in order to collect real-time data from the shop floor. MES receives data from the facilities and stores it on the MES data server. Optional units can exchange internal data with other units using DFA in a B2B method. The CaPU's operating system is responsible for gathering capabilities that are surpassed by DFAs and making them accessible to DFAs that want them, and the CaPU's operating system is the network's core unit. The CPOS is informed about surpassing capabilities on a general, technical, and awareness level. Each unit contains another peer of the "BlockTrust" network. They both play the same game.

BlockTrust Network

The core components of the BlockTrust network are depicted in Fig. 3. For simplicity, just three peers of DFU1, DFU2, and DFU3 are displayed as Peer1, Peer2, and Peer3, respectively. Each peer is equipped with one or more smart contracts and ledgers. The system utilises channels (e.g. channel1, channel2) to link DFA to peers and ultimately to ledgers. Figure 4 depicts the ledger's update and query procedures. As seen in the image, a DFA is necessary to transmit a transaction offer to the proper channel. Consensus is achieved through three distinct steps: verifying the transaction, organising it into a block, and committing the block to the ledger. On a DFA, a transaction is formed by sending a proposal to the preparers and approvers. The approvers then simulate the proposed transaction, and the reading set constructs their sets. After signing and returning the authorization peer read set to DFA, it is

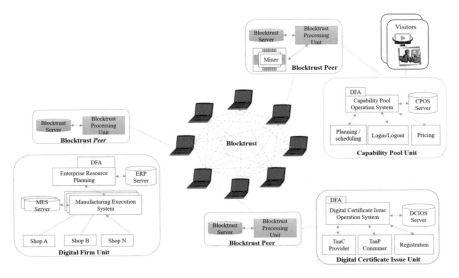

Fig. 2 CM framework embedded with BlockTrust [15]

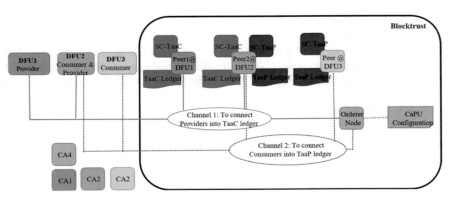

Fig. 3 Internal components of a BlockTrust network [15]

used in further phases of the transaction reconciliation cycle. DFA then sends the read set and its associated approved transaction to the order service. The miner obtains the data from the read set and the allowed trust score transaction, organises it into a block, and distributes it to all committed peers. Finally, the commit peer tells the DFA asynchronously of the transaction's success or failure.

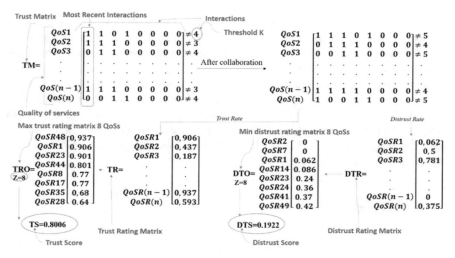

Fig. 4 Trust matrix, trust rating, and trust score [15]

Ledger Structure and Trust Scoring Method

Under contrast to P2P networks, with broad file sharing in blockchain trust, the quality of various services must be controlled. The query-response design of first-generation 2P2 networks, on the other hand, is protocol-for-all. The columns indicate the feedback values as a confidence matrix, whilst the rows represent each QoS. Selçuk et al. 2014's technique is utilised to extract mistrust and trust ratings, as it is a well-known procedure for assessing trust in query-response P2P networks. As seen in Fig. 4, the confidence matrix is a matrix with M columns and N rows. Columns are binary vectors consisting of rows of M bits (e.g., 8) each reflecting a certain sort of service quality (e.g., pricing, quality). 0 indicates discontent, whereas 1 indicates contentment. Each line is preceded by an integer (K) indicating the number of significant bits in that line. After an interaction at the most important bit, the satisfaction/dissatisfaction outcomes are written, moving the available bits to the right.

$$\left(e.g. \frac{\text{Trust Vector: } 11010000}{\text{\#of significant bits : 4}} \xrightarrow{\text{After collaboration}} \frac{\text{Trust Vector: } 11101000}{\text{\#of significant bits : 5}} \right)$$

A row in the confidence The K-bit matrix is divided by 2 K to provide a row in the confidence rating matrix for the range (0, 1). Additionally, the confidence rating matrix A is generated from the confidence matrix's complement in order to convert a confidence matrix to a "confidence rating matrix".

$$\left(e.g. \frac{\text{Trust Vector : } 11101000}{\text{\# of significant bits : 5}} \Rightarrow \frac{\text{To} \quad \text{Trust rating} = \frac{(11101000)2}{2^5} = 0.90625}{\Rightarrow \text{Distrust rating} = \frac{(0\times0000\times0)2}{2^5} = 0.0625} \right).$$

The Z threshold establishes the number of rows (i.e., service quality) to evaluate in a matrix of distrust and confidence ratings. Confidence/distrust grades are initially ordered according to their level of confidence using the minimum-distrust-maximum-confidence criteria. Following the selection of a Z-rank, the trust and mistrust score of the company are calculated as the average of its trust and distrust scores.

The internal interaction between the framework's units may be classified into two broad categories. The initial set of interactions consists of those necessary to add a new member to the network and those necessary to locate a good candidate for a company. The second kind of interactions is those that involve providing trust transactions to the collaboration's nodes.

Interaction Before Collaboration

Issuing a Digital Certificate for a New User

A new business desiring to join the ecosystem as a DFU must have a DC. DC provides the firm with two kinds of information, including the firm's first trust scores and business identification. An application firm (DFA@1) must furnish the CPOS on the CaPU with its general information, technical capabilities, and a history of services received/provided in the past, as seen in the image. The CPOS gathers all of the applicant's information and forwards it to the DCIOS, which contacts the DCIU to seek a DC for the business. The registrar at DCIU immediately utilises this information and develops a DC claim proposal for the applicant based on it. The registration forwards this offer to the DCIO. After a corporation is designated as a provider or a customer in DCIOS, its previous interactions are forwarded to the TaaC and TaaP units for scoring. The TaaC and TaaP units determine the firm's trust scores based on the information given by the applicant. DCIO returns the applicant's points gained, and this unit grants the applicant a DC. Finally, DC is referred to as the applicant and block trust via the DCIU and the CPOS's DFA (Figs. 5, 6 and 7).

Request, Receive a Capability from the Pool

This DFU can request a service from the pool by supplying the qualities of capability, an acceptable price, and adequate TaaP points. The CPOS analyses the request received and returns a list of matching capabilities accessible on the CPOS server. As soon as information is received in CPOS, this unit gives the requester with a list of completed capabilities, including availability, price, and TaaP points. Creates a list of customer DFA responses based on the CPOS provider DFA responses. The consumer may make an order by accepting the DFA list. When the list is approved, a contract is delivered to the provider and consumer DFAs. Figure 8 illustrates the sequence diagram of the interaction between units required to add a new capacity to the pool. As seen in the picture, a DFA must utilise DC to access the system through CPOS and login/logout units. Additionally, this DFA should include complete technical information to the CPOS on the capability and availability supplied, as well

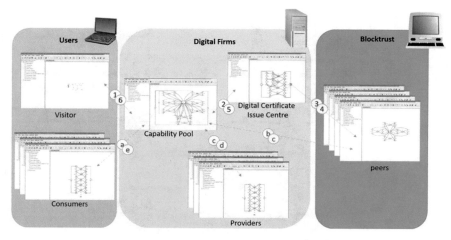

Fig. 5 Interaction amongst the units and BlockTrust [15]

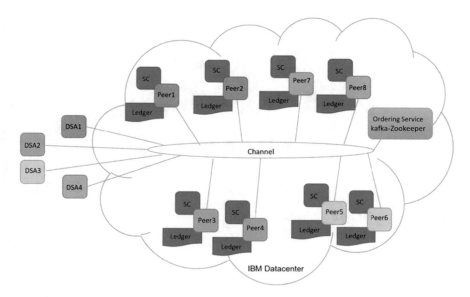

Fig. 6 Case BlockTrust configuration [15]

as price information. Once CPOS receives a response configuration, it will send a confirmation message to DFA, indicating that the capability is now accessible to customers.

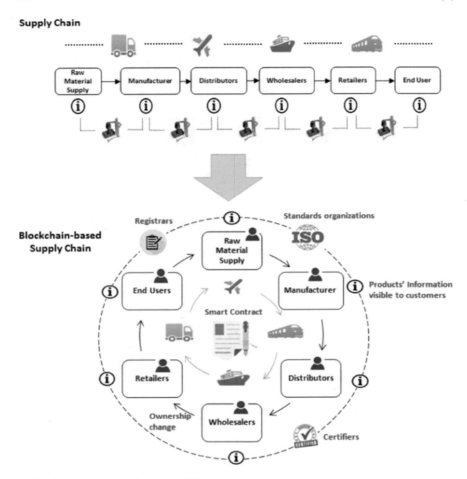

Fig. 7 Supply chain transformation [16]

Interaction After Collaboration

Trust Scores for a Completed Interaction
CPOS solicits TaaC and TaaP transaction proposals from provider and consumer DFAs. The provider DFA creates a TaaC transaction proposal and transmits it to the local blockchain trust peer. After the local peer verifies the DFA's signature, it invokes the chaincode and creates a RW set. Provides local peer-to-peer approval of the project. The authorised transaction is forwarded to DFA by the local peer. DFA collects approvals and transmits the authorised transaction to the orderer via the local peer. The ordering peer then applies the consensus procedure to the authorised transaction, resulting in the creation of a transaction block. The block is then sent to

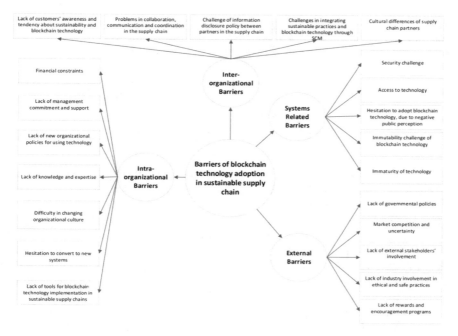

Fig. 8 Barriers of blockchain technology adoption in sustainable supply chain [16]

the node through the transaction ordered by the local peer. The local peer requests that the transaction block be written to the ledger by a committed peer. The CPOS will notify the contractor once the score is complete.

3.1.4 Implementation

As seen in Fig. 5, five distinct types of models were constructed and connected together: visitor, CaPU, consumer company, block trust, and DCIU. For each of these businesses, a trust matrix consisting of sixteen columns (to reflect history interactions) and one row (to indicate service quality) is constructed, and the block is placed in trust. The DCIU generates the trust matrix for the visitor by assigning random integers. By manipulating token flows on and across models, we may analyse interactions inside and between units.

Two test scenarios are utilised to evaluate model interactions: a. the consumer's interaction with the CapU to read a confidence score; and b. the visiting firm's interaction with the CaPU, DCIU, and DCIU to write a trust score. In the first test scenario, the CapPU presents a list of available capabilities to the customer, along with the vendors' trust scores, price, and useable capacity. The consumer then requests a capability from the CaPU. In the second test scenario, a visiting firm requests that the CaPU joins the system by giving general, technical, and interaction information about

Table 1 The used configurations for measuring the performance

Number of channels	1
StateDb database	GoLevelDB
Peer resource	32 vCPU, 3Gbp link
Endorsement policy	OR {AND(a,b); AND(a,c); AND(a,d); AND (b,c); AND(c,d)}
Block size	10, 30, 50, 70
Transaction arrival rate	25, 50, 75, 100, 125, 150, 175

the organisation (three tokens). The CapU contacts the DCIU to provide the business a certificate. Finally, the DCIU issues a token to the visiting firm to indicate that the interactions have concluded. Performance evaluation of BlockTrusts as seen in Fig. 6, the case BlockTrust consists of an orderer node (Kafka-Zookeeper) and four digital businesses, each with two peers. All peers and orderers are implemented in the IBM SoftLayer data centre as X86-64 virtual machines. Four DFAs are utilised to produce the load, each with 54vCPUs and 128 GB of memory. The nodes are linked to the data centre's 3Gbps network. The developed block trust's detailed characteristics are listed in Table 1. The network makes use of GoLevelDB, StateDb, and GoLevelDB. The first step is to investigate the influence of received transaction speed and block size (number of transactions in a block) on throughput and delay. After that, the effect of the number of QoS on throughput and latency is investigated. There have been created configurations that take into account all conceivable block sizes and transaction arrival rates. Six numbers per second were chosen as the transaction arrival rate: 25, 50, 75, 100, 125, 150, 175. Four block sizes were chosen as the size of the ledger blocks: 10, 30, 50, and 70 transactions per block. As a warm-up, each setup component is performed for 100,000 s, followed by an examination of the network's throughput and latency. The effect of received transaction and block size on throughput is shown in Table 2. As seen in the table, efficiency grows linearly with the speed of the arrival processes in all configurations.

The delay of the block trust network is presented in Tables 3 and 4 for each setup. The delay of arrival configurations is continuously growing from 25 to 125 Tx/s. However, in setups with 150 and 175 Tx/s arrivals, the network's latency increases

Table 2 Impact of block size and transaction receive rate on throughput

Throughput (Tx/s)	10 TX in a block	30 TX in a block	50 TX in a block	70 TX in a block
25	22	25	28	30
50	51	54	56	57
75	74	77	79	80
100	98	100	113	105
125	122	123	125	130
150	137	138	139	140
175	139	140	140	141

Table 3 Impact of block size and received transaction rate on latency

Latency (ms)	10 TX in a block	30 TX in a block	50 TX in a block	70 TX in a block
25 Tx/s	400	890	960	974
50 Tx/s	384	759	995	1022
75 Tx/s	368	689	755	1248
100 Tx/s	344	687	769	1301
125 Tx/s	310	694	761	1422
150 Tx/s	15,589	13,484	12,326	10,145
175 Tx/s	44,481	41,801	37,489	33,786

significantly. 125Tx/s can be deemed the optimum acceptable arrival rate for the case block trust network amongst the configurations evaluated. We constructed new trust matrix configurations with 5, 10, 15, 20, 25 columns to investigate the influence of transaction size on the latency and throughput of block-confidence, and we examined several previously existing setups employing confidence matrices. The configurations 125&70; 100&70; 125&50; and 100&50 (transaction arrival rate and transactions per block) are utilised. Table 3 summarises the throughput and latency of the setups that were evaluated. As illustrated in the table, 125 and 70 configuration delays are feasible in a confidence matrix with a maximum of 15 columns. The 100&70 configuration may be applied to up to 20 columns in the confidence matrix, whilst the 125&50 and 100&50 configurations can be used to up to 25 columns.

3.2 Supply Chain Application

3.2.1 Blockchain-Based Supply Chain

The capability of blockchain to secure the trustworthiness, traceability, and validity of information, as well as the potential of smart contractual agreements to create a trustless environment, all lead to a substantial rethinking of supply chains and supply chain management. This section will examine the value proposition of blockchain technology and its application to the goods and manufacturing supply chain, with an emphasis on the article blockchain technology and its relationships to sustainable supply chain management by Sara Saberi, Mahtab Kouhizadeh, Joseph Sarkis, and Lejia Shen.

The mechanisms through which blockchain technology operates inside the supply chain are still being evaluated and improved. As a result, there are several ways to use blockchain technology into supply chain management.

Despite the fact that financial cryptocurrencies can operate on a public network, the supply chain may require a private network with a restricted number of members.

Table 4 Impact of trust matrix size on throughput and latency

# of column in a trust matrix	125 Tx/s &70 Tx in block		100 Tx/s &70 Tx in a block		125 Tx/s &50 Tx in a block		100 Tx/s &50 Tx in a block	
	Throughput	Latency	Throughput	Latency	Throughput	Latency	Throughput	Latency
1 column	130	1400	105	1300	125	760	113	760
5 columns	128	1413	104	1380	121	883	111	840
10 columns	126	1440	98	1440	112	963	107	867
15 columns	124	1476	91	1844	99	1380	101	907
20 columns	21	10,344	84	2340	91	1888	94	1386
25 columns	14	44,810	23	28,480	21	32,475	82	3404

Figure 7, which is an excerpt from the accompanying article, depicts a generic transformation graphic from a traditional supply chain to one that is based on blockchain technology.

Four prominent stakeholders perform critical responsibilities in blockchain-based supply networks, some of which are not evident in traditional supply chains. Registrars, who assign unique IDs to network actors. Standards groups that define systems such as Fairtrade for environmentally friendly supply chains or blockchain rules and technical standards. Certifiers are the organisations that give certifications to anyone seeking to join a supply chain network. Actors who, in order to sustain system confidence, must be certified by a recognised auditor or certifier, such as manufacturers, retailers, and customers [16].

Additionally, this new blockchain-based supply chain network supports product and material movements. To have immediate access to product profiles, each product may have a distinct digital presence. Security measures can be used to restrict access to a product, enabling access to the product's information or the product itself to those who possess the requisite digital keys. With that unique digital key, the state of the product, the type of product, and the standards that will be implemented for the product are all accessible.

The manner in which a product is "owned" or transferred by a particular actor is an intriguing structural and flow management characteristic. Permission for actors to add new information to a product's profile or to enter into a transaction with another party will almost probably be a critical regulation, necessitating smart contract agreements and consensus in order to get permission.

Before transferring (or selling) a product to another actor, both parties may sign a digital contract or satisfy a smart contract condition. Once all parties have fulfilled their contractual obligations and processes, transaction information is updated on the blockchain ledger. When a modification is made, the system updates the records of data transactions automatically [17].

As a result, the blockchain eliminates the need for a centralised trusted entity to operate and maintain the system, allowing consumers to check the continuous chain of custody and transactions from raw materials to ultimate sale. As transactions take place across these several blockchain information dimensions, this data is stored in verifiable ledgers.

Smart contracts, which are stored on the blockchain as written rules, may be used to control how network participants interact with one another and with the system as a whole. Smart contracts have an effect on the interchange of network data and the continuous development of supply chain processes. Digital verification is used by certifiers and standards bodies to certify actor profiles and goods, for example.

Each actor and object on the network have a unique digital profile that comprises information such as a description, a location, certifications, and product associations. Each supply chain member may register critical information about a given product and its present state on the blockchain network [16].

3.2.2 The Barriers Infront of Implementation

The barriers were determined based on the literature related to supply chain information systems, sustainabile supply chains, and blockchain technology. The opinions of experts were also solicited in order to further validate the list of roadblocks. The barriers are summarised and categorised into four categoriesin the Fig. 8: intra-organisational barriers, inter-organisational barriers, system-related barriers, and external barriers, which take into account an organisation's internal and external limitations in adopting a new technology.

4 Conclusion

Blockchain technology offers a plethora of applications and is a quickly expanding and fluid industry. As a result, it is critical to stay current and grasp the most recent advancements in the development and application of blockchain.

Blockchain technology is still in its infancy. The emergence of cryptocurrencies and the development of blockchain technology have established a new paradigm for decentralised security and trust in untrusted systems via the consensus process.

We examined the use of blockchain in manufacturing from the data, network, and application levels, concentrating on the operations or implementations of each, and explored the characteristics given by blockchain.

We have observed appropriate research across a variety of applications and noted that the continued and immutable record provided by blockchain enables permanent trust and auditing of transactions and is ideally suited to the distributed nature of IoT, cloud, and edge computing, as well as many other fields.

References

1. Kuo TT, Ohno-Machado L (2018) Modelchain: decentralized privacy-preserving healthcare predictive modeling framework on private blockchain networks. arXiv preprint
2. Ghasempouri SA, Ladani BT (2019) Modeling trust and reputation systems in hostile environments. Futur Gener Comput Syst 99:571–592
3. Barenji RV (2013) Towards a capability-based decision support system for a manufacturing shop. In: Working conference on virtual enterprises. Springer, Berlin, pp 220–227
4. Li T, He T, Wang Z, Zhang Y (2020) SDF-GA: a service domain feature-oriented approach for manufacturing cloud service composition. J Intell Manuf 31(3):681–702
5. Lu Y, Wang H, Xu X (2019) ManuService ontology: a product data model for service-oriented business interactions in a cloud manufacturing environment. J Intell Manuf 30(1):317–334
6. Yuan M, Cai X, Zhou Z, Sun C, Gu W, Huang J (2019) Dynamic service resources scheduling method in cloud manufacturing environment. Int J Prod Res Ahead-of-Print 4:1–18
7. Barenji RV, Hashemipour M, Guerra-Zubiaga DA (2015) A framework for modelling enterprise competencies: from theory to practice in enterprise architecture. Int J Comput Integr Manuf 28(8):791–810

8. D'Aniello G, De Falco M, Mastrandrea N (2020) Designing a multi-agent system architecture for managing distributed operations within cloud manufacturing. Evol Intell 1–8. (Ahead-of-Print)
9. Wang Y, Wang S, Yang B, Gao B, Wang S (2020) An effective adaptive adjustment method for service composition exception handling in cloud manufacturing. J Intell Manuf 1:1–17
10. Yi H (2020) A post-quantum secure communication system for cloud manufacturing safety. J Intell Manuf 17:1–10
11. Zhang Y, Xi D, Yang H, Tao F, Wang Z (2019) Cloud manufacturing based service encapsulation and optimal configuration method for injection molding machine. J Intell Manuf 30(7):2681–2699
12. Viriyasitavat W, Da Xu L, Bi Z, Sapsomboon A (2018) Block-chain-based business process management (BPM) framework for service composition in industry 4.0. J Intell Manuf 31:1737–1748
13. Nakamoto S (2019) Bitcoin: a peer-to-peer electronic cash system. Manubot, pp 1–24
14. Valdeolmillos D, Mezquita Y, González-Briones A, Prieto J, Corchado JM (2019) BCT technology: a review of the current challenges of cryptocurrency. In: International congress on BCT and applications. Springer, Cham, pp 153–160
15. Barenji RV (2021) A blockchain technology based trust system for cloud manufacturing. J Intell Manuf
16. Saberi SK (2019) Blockchain technology and its relationships to sustainable supply chain management. Int J Prod Res 2117–2135
17. Abeyratne SA (2016) Blockchain ready manufacturing supply chain using distributed ledger. Int J Res Eng Technol 1–10

Virtual Manufacturing, Technologies, and Applications

Şeyda Nur Börü and Senem Masat

1 Introduction

With new technologies that come into our lives every day, our understanding of production and consumption is changing over time. It is possible to say that the use of virtual reality technology in manufacturing processes has become widespread. With the implementation of virtual reality technology in production, the literature is also referred to as virtual manufacturing. In this section, we will first talk about the use of virtual manufacturing. After we define the concept of virtual manufacturing, we will talk about the paradigms of virtual manufacturing such as design centered, production centered, and control centered. After we describe the definition and paradigms of virtual manufacturing and the structure of VM, we will talk about the applications of virtual reality technology in manufacturing and different areas. Virtual reality technology is also one of the rapidly growing and emerging technologies. It is possible to see its industrial use as well as its personal use. In this chapter, we will also be talking about augmented reality and mixed reality concepts that have entered our lives through virtual reality technology. In the final section, we will address the advantages, disadvantages, and challenges of this technology.

2 Definition of Virtual Manufacturing

Virtual manufacturing is the simulation of possible problems with the functionality, manufacturability, or applicability of a product or design in a virtual environment

Ş. N. Börü (✉) · S. Masat
Department of Industrial Engineering, Hacettepe University, Beytepe Campus, 06800 Ankara, Turkey
e-mail: senem.mst.94@gmail.com

before actual production takes place, and the manufacturing process is completed performed in the same environment [1].

So, why do enterprises need virtual manufacturing? At what points does virtual manufacturing contribute to businesses? The goals of commercial enterprises are to respond quickly to the demands of the market, to achieve low cost and high quality, to provide the expected service opportunities, and to make all of these customer oriented. Considering what difficulties enterprises face in their production processes, the following items can be listed: complex production environments; product design dynamics; scheduling change; difficulty in evaluating advantages and disadvantages before commissioning when a new product is designed; difficulty in accessing actual production information at production design stages; it consists in decoupling the relationship between design and production to ensure that the enterprise receives maximum benefits. Virtual production provides an organization with the ability to analyze the manufacturability of a part or product, evaluate and verify production processes and machines, and train managers, operators, and technicians on production systems. Virtual manufacturing offers solutions to all these challenges [2].

3   Paradigms of Virtual Manufacturing

There are three main subcategories of virtual manufacturing: First, the design-centric VM provides engineers and designers with information about the production process so that they can optimize products for their production purposes or learn how production problems can affect product design. They can also save money and time by testing 3D product models and processes instead of building physical prototypes. Secondly, the production-centric VM simulates production processes so that they can be tested and optimized. Design for assembly uses production-based simulations to optimize product and process design for a specific production goal, such as quality, lean applications, and/or flexibility. The last is the control-centric VM, which simulates the controls used to run the actual production processes. It uses machine control models to ensure process optimization during the actual production cycle. Theoretically, design-centric virtual production supports production information to the designer at the design stage. Manufacturing-centric virtual manufacturing uses simulation technology to optimize the manufacturing process while the production planning phase. The virtual production control center uses a machine control base in simulation, the goal of which is process optimization during actual production [2] (Fig. 1).

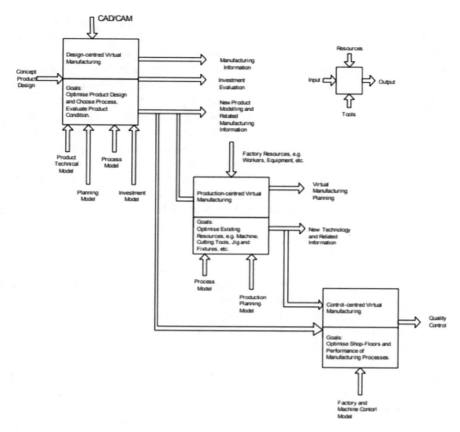

Fig. 1 Relationships among three types of virtual manufacturing [2]

3.1 Virtual Manufacturing Technology

Product performance, manufacturability, and assembly capacity may all be predicted using virtual manufacturing technologies. Thus, manufacturing becomes more efficient, economical, and flexible in its work organization: Premises and workshops are arranged in a more reasonable and effective manner; expenses are reduced; product life cycles are extended; product quality and design are improved, and processes are optimized [2].

3.1.1 Virtual Design

Modern product design is a method that takes the market into account and appeals to buyers, and hence, it is applicable to all markets. It has a multi-level design and an intelligent design. Designed to adapt swiftly to Sunday changes, conserve resources, and safeguard the environment. As a result, the design technique emphasizes cooperation, synergy, and concurrent design. The structure, function, and cost of the product are determined by the design and production of the product. Consider virtual product creation as a form of design-driven virtual manufacturing technology. It generates a digital model rather than a physical one and hence supersedes the original product design. Models are accurate the first time, assisting in the reduction of material use. Virtual design enables interdisciplinary collaboration between designers and engineers to complete tasks such as product design and possible adjustments, production, and testing [3]. Virtual design is employed not just at the prototyping stage but also throughout the manufacturing and sales processes. Virtual design technology is extensively applicable in a variety of fields, including fast design and prototyping, assembly design, manufacturing design, design for disassembly, parallel machining for product design, and staff training and product servicing [2].

3.1.2   Virtual Machining

Virtual processing is the process of altering a product's geometric features and physical performance in a virtual environment. Cutting process simulation, welding process simulation, pressing process simulation, and casting process simulation are all examples of machining process simulation. It includes. The fundamental concept of virtual machining is the use of information technology, modeling, and simulation of the whole production process on three levels. At the technical process level, emphasis is placed on precise and trustworthy data support and modeling. At the level of the actual machining simulation process, it enables a speedy and cost-effective evaluation at the design stage of the technical process. At the production system level, emphasis is placed on efficient performance evaluation, modeling the whole product life cycle, simulating enterprise virtual operations, and utilizing risk assessment to alter and improve the system [2] (Fig. 2).

3.1.3   Virtual Assembly

Virtual assembly is accomplished by creating an intuitionistic visible dynamic model on a computer and inspecting each component assembly in a virtual environment using interferential inspection [4, 5]. The advantage is that it can quickly identify and correct design errors, hence increasing design efficiency and lowering costs. The quotation approach is used in virtual assembly. It does not add all components to the assembly model; rather, it recalls their positions within the model. When components are necessary, they are forwarded to the assembly model for assembly. As a result, these operations conserve hard drive and memory space. The advantage of adopting the quote technique is that the assembly model will only contain information about the most recent component assembly. Following alteration, the assembly model may be automatically updated. As a result, it can help minimize workload. Another advantage of the quote approach is that it enables concurrent engineering to be

Fig. 2 Virtual assembly
modeling process [2]

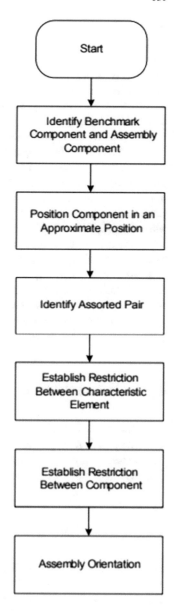

supported technically. Each component referenced by the assembly model can be
saved as a file on a computer. As a result, it is simply controlled and transported
through networks, and it enables cooperation. The virtual assembly supports two
types of assembly: top to bottom and bottom to top. Designers can choose from a
variety of virtual assembly models based on the various product attributes [2].

3.1.4 Virtual Test

Testing, measuring, and inspecting are all effective technical tools for ensuring the quality of a product. It entails executing checks throughout the manufacturing process, including product design and manufacture, assembly performance analysis, and quality control, to guarantee that quality control and normal production are carried out. It completes inspection, testing, and monitoring operations in a physical simulation environment by utilizing different elements of the virtual product design model. Virtual manufacturing check items include virtual component geometry dimensional tolerances, virtual component geometry dimensional tolerances, virtual component geometry dimensional tolerances, virtual component surface roughness, and virtual manufacturing process parameters. Additionally, it encompasses product performance testing, virtual product movement, stress and strain distribution testing, vibration impact testing, fault and noise impact testing, and diagnostics. The virtual test system is composed of virtual test subsystems that simulate the product's design, virtual machining, virtual assembly, and performance. Virtual testing data may be utilized to refine and improve design and process parameters throughout the virtual manufacturing process. At the design stage, virtual designers can conduct product testing. Engineers of test and design can collaborate. As a result, it is capable of avoiding and resolving test issues caused by design changes. Virtual testing has the potential to drastically reduce the time required to design a new product [2] (Fig. 3).

4   Technical Support for Virtual Manufacturing

These are some of the well known technologies in virtual manufacturing: computer-aided design (CAD), product lifecycle management (PLM) and simulation software, 3D modeling, rapid prototyping, virtual reality, high-speed networking services [6] (Fig. 4).

4.1   Virtual Reality

In recent years, virtual reality (VR) technology has gained increasing interest in a range of businesses. Virtual reality technology, which is always changing and expanding, provides the opportunity to improve a variety of applications in a variety of sectors. The industrial industry is leading the charge in developing applications for virtual reality. Additional research has revealed that considerable effort is being invested in virtual reality applications in manufacturing. It reveals that it has significant benefits for the regions' progress especially, for multinational manufacturing businesses with personnel and operations in many locations [7].

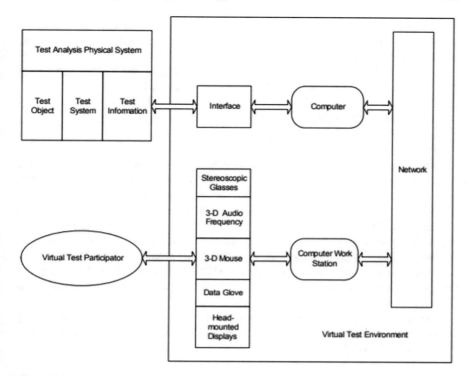

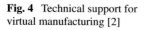

Fig. 3 Virtual test environment [2]

Fig. 4 Technical support for
virtual manufacturing [2]

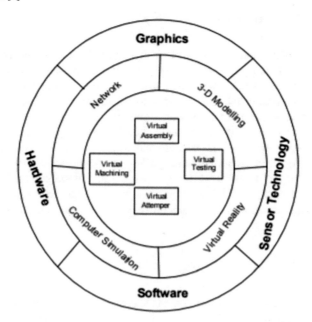

The first VR system was successfully developed using an HMD that delivers a stereoscopic 3D vision to the user and is coupled to a sensor device that records the user's head movement [7].

Since then, further study has been conducted in this sector. Korves and Loftus classified the VR system into the following configurations [8]:

Desktop computers.

Massive projection system.

CAVE technology that is completely immersive.

DMH-based submersible VR system.

Due to recent advancements in the hardware and software of VR devices, the immersive VR system with HMD is becoming increasingly relevant in a wide variety of industries, and as a result, it has regained significant interest in academia and industry during the last decade. Numerous researches on various aspects of virtual reality's incorporation into the manufacturing process exist in this burgeoning topic. Numerous individuals have showed how virtual reality may be used to augment existing manufacturing applications. For instance, Wiendahl and Harm-schristian demonstrated that immersive virtual reality is a critical tool for collaborative production planning, much more so when several user viewpoints are depicted [9].

Menck observes that a collaboration planner powered by virtual reality may expand communication and cooperation beyond traditional organizational borders so reducing job complexity and increasing work efficiency [10].

4.2 Augmented Reality

Augmented reality (AR) is a similar idea, but instead of isolating reality from computer-generated information, it incorporates computer-generated content into the real-world experience, allowing the two to be enjoyed simultaneously together. According to Milgram and Kishino, augmented reality exists on a continuum between the actual and virtual worlds, resulting in "mixed reality". Augmented reality (AR) is a similar idea, but instead of isolating reality, it incorporates or embeds computer-generated material into the real-world experience, allowing both to be experienced. AR decodes numeric data pertaining to physical components [11]. Pokémon GO* is maybe the most well-known example. Augmented reality maintains a focus on the physical world, but augments it with digital elements, enveloping it in a new layer of perception, and augmenting reality or your environment [12].

4.3 Mixed Reality

MR uses digital components to deconstruct the actual environment. You may interact with and traverse actual and virtual objects and surroundings using next-generation mixed reality image and detecting technologies. Mixed reality enables you to observe and immerse yourself in the world in which you are currently located, even if you interact with the virtual environment with your hands without removing your helmet. It enables you to wander around in the actual world with one foot (or one hand) and the fantasy world with the other. It deconstructs reality and imagination's fundamental principles, providing an immersive experience that may revolutionize the way you play or work today [12].

4.4 X-Reality

Virtual reality (VR) immerses users in an entirely virtual environment; augmented reality (AR) produces a virtual content overlay that adapts to the environment; and mixed reality (MR) generates virtual things that interact with the real world. XR uses a single phrase to denote the division of three worlds (AR, VR, and MR) [12].

5 Applications of VR

Virtual reality is a technology with a wide range of applications, and we see that it is used in different areas today: health, tourism, aerospace, etc.

5.1 Health Care

Augmented reality and virtual reality applications are used in a wide range of health care. It is possible to see many applications from employee training to disease treatment. With virtual reality technology used in doctors' and nurses' training, doctors have the opportunity to perform surgery in virtual environments, thus preventing mistakes and losses as they have a chance to predict complications during the intervention. In addition, virtual reality applications are thought to remove people from the real world, so patients are also perceived to be away from the pain and pain they feel during intervention when they use VR glasses. In the treatment of phobia diseases, the situation that the person is afraid of (fear of elevators, fear of heights, etc.) he needs to face, but in real life, it is not possible; the virtual reality applications used for this purpose allow patients to experience and overcome their fears in a safe and virtual environment. Apart from these applications, there are different applications such as early diagnosis of Alzheimer's disease, post-accident rehabilitation [13].

5.2 Education

We see that virtual reality technology is also widely used in education. The purpose of virtual reality technology is to experience it. The information is more permanent because students experience courses such as history, geography, and so on. Students are more interested in learning as they use new technology. Because traditional education processes could not support abstract thinking, students were forced to do abstract thinking, but with virtual reality technology, students can also learn abstract concepts. Because of its use in fields such as art, entertainment, this technology extends the imagination of students and increases creativity. Virtual reality applications also allow students to collaborate in groups [13].

6 Virtual Manufacturing System Applications

Despite the fact that there are different definitions, virtual manufacturing is widely employed in the military, industry, research, and education sectors, among others. It has been effectively employed in a wide number of fields. Virtual manufacturing offers limitless possibilities in manufacturing design and manufacturing process applications. It has the ability to considerably improve the technological level of the product. Virtual manufacturing has considerably increased manufacturing flexibility while also cutting costs. The following examples illustrate advanced virtual applications in one or more design and manufacturing fields. [2] When it comes to applications, there are a few critical points to keep in mind: Virtual automobile design includes product layout design, product assembly simulation, product machining process simulation, and rapid prototyping [14].

The first industry that comes to mind is FORD MOTOR. Several individuals from several automobile manufacturers discussed their experiences with driver visibility analysis. Due to the fact that the majority of cars have three sets of pillars, it is critical to understand how their size and location impact the driver's view of the outside world. While larger pillars may result in safer automobiles, they sacrifice driver visibility. We noticed several instances when designers sat in a virtual automobile and organically walked around the surroundings, leveraging HMDs or CAVEs to get a genuine feeling of the available visibility [14].

Engineers at General Motors' Design Lab are investigating the effect of veiling glare from instrument panels on the driver's glass. Due to advancements in lighting algorithms, exact computation and production of light reflections are now possible. Sitting in a real vehicle seat in their five-walled CAVE is an intense sensory experience. The combination of the high-resolution monitor and the lighting simulation created an immersive experience. This method enables designers to have a better understanding of how the instrument panel influences driver visibility during night driving [14].

Additionally, it was quite common to measure visibility in terms of motion and interaction. To better grasp operator perspectives during a transmission docking activity, Ford Ergonomics Lab ergonomic specialists used a head-mounted display (HMD). Engineers sought to reduce the length of the studs (visual guides) needed to dock the transmission in a single spot. On the other hand, shorter studs obscure them. Engineers used virtual reality to verify that the studs remained visible during the docking operation [14].

Ford Motor Company's ergonomic specialists are leveraging virtual reality to establish design standards for the maximum assembly force authorized when fitting various hoses. Ergonomic engineers calculated the forces necessary to attach hoses in various human positions using an HMD, physical props, and force sensors. They used the data gathered during their virtual reality experience to develop design requirements for external suppliers pertaining to the maximum force necessary for installation. In this scenario, virtual reality enables personnel of varying heights and strengths to securely do assembly tasks [14].

Ergonomic engineers assess the reachability of door knobs within a car buck using a massive stereo power will play in one of Case New Holland's VR laboratories. Several alternative door handle placements were eventually eliminated during one design iteration due to their unpleasant and hazardous position for the driver. Due of the big wall architecture, Case New Holland may also use VR to examine the reachability of an exterior door handle from a ground-level stance for certain of their heavier merchandise [14].

PSA Peugeot Citroen designers use a three-sided CAVE to investigate alternative control positions inside vehicle designs. Control and instrument placement within the overall architecture has a considerable influence on the overall feel of the interior [14].

Quick prototyping and rapid manufacturing are terms that refer to a series of techniques that begin with a CAD model and proceed layer by layer. These methods, which may be applied as powder, paste, or liquid, can be used on polymers, ceramics, and metals [14].

7   Advantages, Disadvantages, Challenges, and Future of VR

7.1   Advantages

We have seen that virtual reality technologies have different uses, of course, and their advantages vary depending on the area in which it is used. And what we come to use in the industry is that we can count its benefits in this way, reducing risk, accelerating R&D processes, preventing manufacturing errors, providing cost benefits, increasing efficiency [15].

7.2   Disadvantages

It is believed that virtual reality technology has created as many threats as it has benefits. Let's talk about some threats, especially in the research of entertainment personal use. Children who met early age with virtual reality technology were observed to be unable to distinguish between the real world and the virtual world. This is a big problem for our future. Other than that, there are studies that these practices for adults are unconsciously harmful. There are reports that an augmented reality game that has been in our lives over the past few years has increased traffic accidents. Other than that, adults have war, adventure, etc. They were identified as showing signs of post-traumatic stress disorder when they played virtual reality games. So, this technology, which is useful in its time, place, and dose, can also damage our life, health, our future [16].

7.3   Challenges

The expansion of this emerging technology is expected to be applied to all areas of life, but it doesn't seem likely to happen quickly. There are some challenges in the proliferation of virtual reality technologies.

7.3.1   Social Rejection

Virtual reality technology is used in many different areas, leading to the outset of traditional methods. This means that social and cultural structures will change in the process ahead. Many societies will resist technology, thinking that this technology will damage the social structure. In addition, users' expectations of security, privacy, and so on are not fully met, as it is a newly emerging technology [17].

7.3.2   Security Privacy

As with the emergence of every new technology, one of the biggest problems encountered in the proliferation of VR technology is security and privacy. This is important for businesses and for personal purposes, to use this technology. Businesses want the technology they integrate into their systems to meet their security expectations and so do other users. There are not enough security standards as there is a new technology, which is a concern for users. The development of standards to reduce or prevent potential risks must be made by law to ensure safety [17].

7.3.3 Regulation/Legal

As a result of the two key concerns we are talking about above, governments can also sanction the production, use, and proliferation of this technology. The development of technology is slowing because some legal constraints are applied [17].

7.3.4 Technical Challenges

New technologies make our lives easier. But, when we look at the infrastructure behind these technologies, we notice that they have important software and hardware. To use and promote virtual reality technology, the background infrastructure must also be robust. The low cost of hardware and software processes will also result in lower prices of products. However, more users will be available when this is possible [17].

7.4 Future

In the bar graph that we present above, you see consumer spending on virtual reality hardware. In addition to its benefits, we see that the challenges and threats that it faces are increasing by the year, and in the years to come, it is possible to extract it from this graph that it will become widespread and become part of our lives (Fig. 5).

8 Conclusion

The benefits of using virtual reality technology, especially in manufacturing processes, are quite high for businesses. It is expected that the concept of virtual manufacturing or technology will evolve over time and that many businesses will integrate this technology into their systems.

As we have seen in the last part, virtual reality technologies are becoming more and more common every day. We talked about the use of this technology in manufacturing processes, the use of it in non-production industries, which we often see in different areas throughout the process. It is possible to predict that the technology will continue to spread rapidly if at least some of the expectations are met, and the challenges in the development and expansion of the technology we are addressing in this chapter. Given the benefits of virtual reality and other technologies in the industry, we expect rapid spread in industries such as manufacturing, healthcare, and education. We anticipate

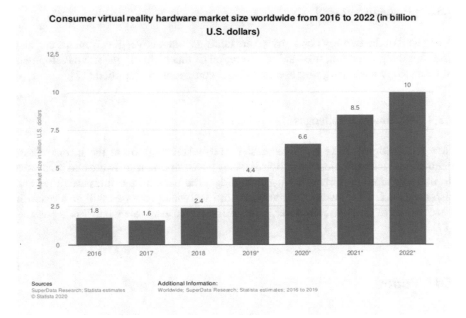

Fig. 5 Virtual reality hardware market through years [18]

that these technologies will soon be part of our lives, as there are disadvantages, but there are more advantages.

References

1. Wiens GJ (1995) An overview of virtual manufacturing. In: Proceedings of the agile manufacturing conference 1995, Albuquerque, NM, pp 1–15
2. Wang Q (2005) Virtual manufacturing and systems. WIT Trans State Art Sci Eng16
3. Rumbaugh J, Blaha M, Premerlani W, Losenson W (1991) Object-oriented modelling and design. Prentice-Hall Inc., New Jersey
4. Jayaram SH, Connacher H, Lyons K (1997) Virtual assembly using virtual reality techniques. Comput Aided Desig 29(8)
5. Lotter B (1989) Manufacturing Assembly Handbook. Butterworths, London, UK
6. https://searcherp.techtarget.com/definition/virtual-manufacturing#:~:text=The%20main%20t echnologies%20used%20in,speed%20networking%20and%20rapid%20prototyping
7. Sutherland IE (1968) A head-mounted three dimensional display. In: Proceedings of the December 9–11, 1968, fall joint computer conference, part I, pp 757–764
8. The application of immersive virtual reality for layout planning of manufacturing cells. Proc Inst Mech Eng Part B J Eng Manuf 213:87–91 (1999)
9. Wiendahl H-P, Harmschristian Fiebig T (2003) Virtual factory design—a new tool for a co-operative planning approach. Int J Comput Integer Manuf 16(7–8):535–540
10. Menck N et al (2012) Collaborative factory planning in virtual reality. Procedia CIRP 3(1):317–322

11. Mann S et al (2018) All reality: virtual, augmented, mixed (x), mediated (x, y), and multimediated reality. arXiv preprint arXiv: 1804.08386
12. https://www.intel.com/content/www/us/en/tech-tips-and-tricks/virtual-reality-vs-augmented-reality.html
13. Syncvr.tech
14. Berg LP, Vance JM (2017) Industry use of virtual reality in product design and manufacturing: a survey. Virtual Reality 21(1):1–17
15. https://www.nsocialtr.com/uretimde-sanal-gerceklik-kullanimi-uretimde-vr-kullanimi.html
16. Virtual reality: ethical challenges and dangers/ben kenwright/2018
17. Augmented reality and its effect on our life/Riya Aggarwal ASET, Abhishek Singhal ASET/2019
18. Barenji AV, Barenji RV, Hashemipour M (2016) Flexible testing platform for employment of RFID-enabled multi-agent system on flexible assembly line. Adv Eng Softw 91:1–11

Digital Twin and Its Applications

Merve Melis Ergün, Ayşegül Kocabay, Yıldız Merve Yesilcimen,
and Merve Turanli Parlaktuna

1 Introduction to Digital Twin

Increasing product requirements and rapidly changing markets require highly precise product developments in ever shorter cycles. Dependencies and boundary conditions of the later production have to be considered already in the early phases of the product design to ensure the product function. This leads to increasing requirements in the product engineering process. Precise designs with increasing complexity and many variables must be derived. This results in increasingly complex product specifications with regard to the quality requirements. Also, the production of new, highly precise products faces the challenge of producing high quality requirements and interacting features cost-effectively. Often, manufacturing processes are already reaching their technological limits. The current developments in information technologies open up great possibilities for support in the product engineering process through increased computing power, new simulation, and analysis tools, as well as connected data. Digital twins of the product or production are already being modeled in the individual domains in order to derive optimal solutions. The increased data availability and traceability of products also allows the modeling of data-driven models using artificial intelligence methods.

M. M. Ergün (✉) · A. Kocabay · Y. M. Yesilcimen · M. Turanli Parlaktuna
Department of Industrial Engineering, Hacettepe University, Beytepe Campus, 06800 Ankara, Turkey
e-mail: mervemelisergun@gmail.com

2   Digital Twin Architecture

2.1   Digital Twin–The Fundamentals

2.1.1   What is a Digital Twin?

The first terminology was given by Grieves in a 2003 presentation and later documented in a white paper setting a foundation for the developments of Digital Twins. The National Aeronautical Space Administration (NASA) released a paper in 2012 entitled "The Digital Twin Paradigm for Future NASA and U.S. Air Force Vehicles", setting a key milestone for defining Digital Twins.

a: NASA 2012 "A Digital Twin is an integrated multi-physics, multiscale, probabilistic simulation of an as-built vehicle or system that uses the best available physical models, sensor updates, fleet history, etc., to mirror the life of its corresponding flying twin" [1].

b: [2] "A digital twin is a computerized model of a physical device or system that represents all functional features and links with the working elements" [2].

c: liu et al. 2018

    "The digital twin is actually a living model of the physical asset or system, which continually adapts to operational changes based on the collected online data and information and can forecast the future of the corresponding physical counterpart." [3]

d: Zheng et al. 2018

    "A Digital Twin is a set of virtual information that fully describes a potential or actual physical production from the micro atomic level to the macro geometrical level." [4]

e: Vrabič et al. 2018

    "A digital twin is a digital representation of a physical item or assembly using integrated simulations and service data. The digital representation holds information from multiple sources across the product life cycle. This information is continuously updated and is visualized in a variety of ways to predict current and future conditions, in both design and operational environments, to enhance decision-making" [5].

f: Madni 2019.

    "A Digital Twin is a virtual instance of a physical system (twin) that is continually updated with the latter's performance, maintenance, and health status data throughout the physical system's life cycle" [6].

There are three crucial characteristics of Digital Twins, which should be included in the definition:

1. The Digital Twin is a virtual dynamic representation of a physical artifact or system,

2. Data is automatically and bidirectionally exchanged between the Digital Twin and the physical system,
3. The Twin entails data of all phases of the entire product lifecycle and is connected to all of them.

Consequently, the following definition was derived in accordance with the industry partner:

A Digital Twin is a virtual dynamic representation of a physical system, which is connected to it over the entire lifecycle for bidirectional data exchange.

The physical twin automatically transfers, among others, data of its behavior, its status, and information on the environment from the real space to the virtual space over the entire product lifecycle, when needed. The virtual twin instead identifies product or process oriented improvements, control demands based on the current situation, or predictions of the near future and sends them back to the real space so the physical product adapts accordingly.

2.1.2 Definition

A digital twin is a digital representation of a physical object, process or service. The digital twin could be a digital copy of an object in the physical world, such as a jet engine or wind farms, or even larger items such as buildings or even entire cities. Digital twin refers to the processes and methods for describing and modeling the properties, behavior, formation process and performance of physical objects using digital technology and can also be called digital twin technology. A digital twin model refers to a virtual model that fully corresponds to and is consistent with real-world physical assets and can simulate their behavior and performance in real-time in a real-time environment.

Along with physical assets, digital twin technology can be used to replicate processes to gather data to predict how they will perform.

A digital twin is, at its core, a computer program that uses real-world data to create simulations that can predict how a product or process will perform. These programs can integrate the Internet of things (Industry 4.0), artificial intelligence and software analytics to improve output.

With the advancement of machine learning and factors such as big data, these virtual models have become a staple in modern engineering to drive innovation and improve performance.

In short, building one can be accomplished by advancing strategic technology trends, preventing costly failures in physical objects, as well as using advanced analytics, monitoring, and forecasting capabilities, testing processes and services [7].

2.1.3   Digital Twin Misconceptions

- Digital Model
  A digital model is defined as a digital version of a pre-existing or planned physical object, there is no automatic data exchange between the physical model and the digital model to accurately describe a digital model. Examples of digital models can be, but are not limited to, building plans, product designs, and development plans. The important defining feature is that there is no automatic form of data exchange between the physical system and the digital model. This means that once the digital model has been created, a change to the physical object has no effect on the digital model. Figure 1 shows a Digital Model.
- Digital Shadow
  A digital shadow is a digital representation of an object that has a one-way flow between the physical and digital object. A change in the state of the physical object leads to a change in the digital object, not vice versa. Figure 1 shows the Digital Shadow.
- Digital Twin
  If data was flowing between an existing physical object and a digital object and was fully integrated in both directions, this would constitute the "Digital Twin" reference. A change to the physical object automatically leads to a change to the digital object and vice versa. Figure 1 shows a Digital Twin.

These three definitions help identify common misconceptions in the literature. However, there are some misconceptions that can be seen, but they are not limited to these specific examples. Among the misconceptions is the misconception that Digital Twins have to be a complete 3-D model of something physical. If we briefly summarize the digitalization systems, if there is a bidirectional manual data flow, it is called a digital model, if there is a one-way automatic data flow from physical to digital, digital shadow, if there is manual data flow from digital to physical, it is called digital shadow, if there is automatic data flow from both sides, it is called DT.

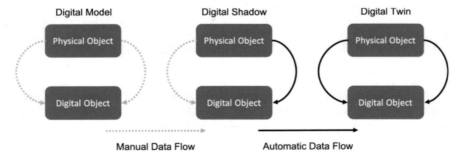

Fig. 1 Digital model, shadow and twin

2.2 Digital Twin Architecture

2.2.1 Basic Features of the Product Digital Twin Model

The product digital twin model has many characteristics including: virtuality, uniqueness, multi-physical, multiscale, hierarchical, integrated, dynamic, super-realistic, computability, probability, and multidisciplinary.

- Virtuality: the product digital twin model is a physical product in digital mapping model, information space is a virtual model, belonging to the information space (or virtual space) and does not belong to the physical space.
- Uniqueness: a physical product corresponds to a product digital twin model.
- Multi-physical: the product digital twin model is based on the physical properties of physical product digital mapping model; It is not only necessary to describe the geometric properties of the physical product (such as shape, size, tolerance.), but also to describe the various physical properties of the physical product, including structural dynamics models, thermodynamic models, stress analysis models, fatigue damage models, and material properties of product composition materials (such as stiffness, strength, hardness, and fatigue strength).
- Multiscale: the product digital twin model not only describes the macroscopic properties of the physical product, such as geometric dimensions, but also the microscopic properties of the physical product, such as the microstructure of the material, the surface roughness, and so on.
- Hierarchical: the different components, parts, etc. that make up the final product can all have their corresponding digital twin models. For example, the aircraft digital twin model includes the rack digital twin model, the flight control system digital twin model, the propulsion control system digital twin model, etc., which is conducive to hierarchical and detailed management of product data and product models, and the progressive realization of the product digital twin model.
- Integrated: the product digital twin model is a multiscale and multilevel integrated model of multiple physical structure models, geometric models, and material models, which is conducive to the rapid simulation and analysis of the product's structural and mechanical properties.
- Dynamic: the product digital twin model will constantly change and improve through the continuous interaction with the product entity during various stages of the whole lifecycle; for example, product manufacturing data (such as test data, the progress data) will be reflected in the digital twin model of the virtual space, and at the same time, based on the digital twin model, can realize the real-time, dynamic and visual monitoring of the manufacturing state, and process of the product.
- Super-realistic: the product digital twin model and the physical product are basically identical in appearance, content, and nature, with high degree of actuality, and can accurately reflect the real state of the physical product.

- Computability: based on the product digital twin model, simulations, calculations, and analysis can be used to simulate and reflect the status and behavior of the corresponding physical product in real-time.
- Probability: the product digital twin model allows computation and simulation using probabilistic statistics.
- Multidisciplinary: the product digital twin model involves the intersection and fusion of multiple disciplines such as computational science, information science, mechanical engineering, electronic science, physics, etc., and has multidiscipline.

2.3 How Does Digital Twin Works?

As mentioned above, digital twins can be created for a wide variety of applications, such as testing a prototype or design, evaluating how a product or process will work under different conditions, and identifying and monitoring their lifecycle.

A digital twin design is made by collecting data and creating computational models to test. This may include an interface between the digital model and a real physical object to send and receive feedback and data in real-time.

- Data
  A digital twin needs data about an object or process to create a virtual model that can represent behaviors or states of a real-world item or procedure. This data may relate to a product's life cycle and may include design features, manufacturing processes, or engineering information. It may also contain production information, including equipment, materials, parts, methods and quality control. Data can also be operational related, such as real-time feedback, historical analysis, and maintenance records. Other data used in the digital twin design may include business data or end-of-life procedures.
- Modeling
  Once the data is collected, it can be used to create computational analytical models to show work effects, predict conditions such as fatigue, and determine behavior. These models can predict actions based on engineering simulations, physics, chemistry, statistics, machine learning, artificial intelligence, business logic, or goals. These models can be demonstrated through 3-D representations and augmented reality modeling to help people understand the findings.
- Linking
  Findings from digital twins can be linked to create an overview, for example, by taking the findings of equipment twins and putting them in a production line twin, then informing a factory-scale digital twin. By using connected digital twins in this way, it is possible to enable smart industrial applications for real-world operational improvements and improvements.

3 Digital Twins Tools and Technologies

The expansion of Internet connectivity into physical devices and ordinary objects has dramatically changed how people interact and communicate in all parts of their lives. Thanks to the Internet, devices can communicate with each other. Device data can be monitored and managed remotely. Known as the Internet of things (IoT), this concept is transforming the way people communicate with physical devices and the environment and is having a significant impact on people's lives [8].

When creating a digital twin of a physical object, companies leverage related products and tools. Shown below in the company-related product/tools order:

Siemens -Siemens PLM Software:

A digital twin is a virtual version of a real product or process that is used to better understand and forecast the original product's functional properties.

In general, there are three sorts of digital twins: product, production, and performance. The term "digital thread" refers to the merging and integration of these three categories as they progress together. Weaving pulls together data from all stages of the product and production life cycle, hence the name thread.

Digital twins of products illustrate how they act in the real-world and provide a virtual-physical relationship. All of this removes the need for several prototypes, cuts product development time in half, increases final product quality, and allows for a far faster reaction to customer feedback.

- Product digital twins of all manufacturing equipment can help to improve output even more. Businesses may minimize costly downtime and even predict when preventative maintenance is required via using data of product and production digital twins. Manufacturing processes become faster, more efficient, and more trustworthy as a result of this steady influx of accurate data.
- Production digital twins are a step taken before a product is manufactured.
- Performance digital twins analyze operational big data created from smart items and facilities. This improves the efficiency of the product as well as the production system [9].

General Electrics-Predix Platform:

The Predix Platform is the IoT edge-to-cloud basis that all GE Digital apps are built on. Predix is designed specifically for processing, managing, and analyzing enormous amounts of machine data. Basic Predix services consists of five categories: Industrial Services, Data Services, Analytical Services, Security Services, and Software/Configuration Services [10].

Microsoft- Azure IoT Hub:

Azure is a cloud application development and infrastructure platform service by Microsoft. This platform, which was launched in 2020, is utilized in a variety of settings (offices, schools, hospitals, banks, stadiums, warehouses, factories, parking lots, streets, parks, etc.). It has real-time monitoring capabilities. It is possible to manage, create, alter, and destroy digital twins using this software (Digital Twins—Modeling and Simulations n.d.).

IBM-IBM Watson IoT Platform:

The IBM Watson IOT Platform was created to generate a virtual bit representation of a physical object or system throughout its existence, allowing for understanding, learning, and reasoning utilizing real-time data (IBM Digital Twin Exchange—FAQ—Saudi Arabia n.d.).

SAP-SAP Leonardo Platform:

To generate a live digital representation or software model of a connected physical device, SAP designed the SAP Leonardo Platform (SAP Digital Twin Software and Technology n.d.).

3.1 Digital Twin Technologies

Lıı et al. Discussed digital twin Technologies in three categories: data related technologies, high-fidelity modeling technologies, and model-based simulation technologies.

- Data Related Technologies:
  The digital twin is built on data. To capture overall data for digital twins, sensors, digital gauges, RFID tags and readers, cameras, 3-D scanners, and other devices must be chosen and interconnected. The data should then be sent in real-time or near-real-time. The data that the digital twin requires, on the other hand, is frequently voluminous, fast, and various, making transfer to the cloud server challenging and expensive. Edge computing is thus a suitable approach for pre-processing acquired data to reduce network load and eliminate data leakage, while 5 G technology enables real-time data transfer. To comprehend the obtained data, data mapping and data fusion are required. XML is the most widely used data mapping technology in the literature.
- High-fidelity modeling Technologies:
  Digital twins are built around a model. Semantic data models and physical models are two types of digital twin models. Artificial intelligence technologies are used to train semantic data models utilizing known inputs and outputs. Physical models necessitate a thorough grasp of mechanical properties and their interrelation- ship. As a result, multi-physics modeling is required for high-fidelity digital twin modeling. Modelica is the most widely used multi-physics modeling language in the literature.

- Model-based simulation Technologies:
  A significant feature of the digital twin is simulation. Digital twin simulation allows a virtual image to communicate bidirectionally with a physical object in real-time [11].

3.2 Product Design Process Based on Digital Twin

Producing a digital twin of a physical product can be done in six steps if the physical object is present. There is no requirement that you follow this order and complete all of the tasks. Furthermore, these steps can be completed at the same time.

The technologies used and how these technologies can be used are exemplified in the steps that can be followed while creating a digital twin of a product.

Step 1: The first stage is to develop a virtual replica of the physical thing. This demonstration can make use of computer-aided design (CAD) and 3-D modeling. Three aspects make up the virtual product: elements, behaviors, and regulations. The product, the user, and the environment, among other things, make up the virtual product model at the element level. It comes with both a geometric and a physical model. At the behavior level, not only the product's and users' behavior is examined, but also the product-user interaction. There are optimization and prediction models at the rule level.

Step 2: Involves analyzing and integrating data from the physical product and the Internet before visualizing it. Direct decision-making necessitates the use of data analytics. Because product data is collected from a variety of sources, it is also vital for uncovering new and fascinating trends. Artificial intelligence techniques (reasoning, problem solving, and so on) can be utilized at this step.

Step 3: In the virtual environment, the product behaviors are mimicked. This step makes use of simulation and virtual reality (VR). Virtual reality has recently been employed in the construction of product prototypes and product design.

Step 4: The physical product is told to perform the actions that have been recommended. The two technological backbones of a digital twin are sensors and actuators. Furthermore, augmented reality (AR) technologies can be employed to reflect some aspects of the virtual product in the actual environment.

Step 5: At this point, real-time, bidirectional, and secure links between the physical and virtual products should be established. Various technologies such as network connectivity, cloud computing, and network security are employed for creating connections. Network connections deliver constant data from the product to the cloud, which powers the virtual product. Bluetooth, QR codes, barcodes, and Wi-Fi are examples of these systems. The product can also be protected, developed, and distributed on the cloud thanks to network technologies. As a result, designers and consumers would have easy access from anywhere with an internet connection. Data security is crucial at this point.

Step 6: Data collected and processed in real-time utilizing sensor and IoT technology. This cycle is repeated, with the obtained data being utilized to develop a more functional digital twin [12].

3.3 The Future, Present, and Past of Digital Twin Technology

One of the key concepts linked with the Industry 4.0 wave is the Digital Twin (DT). Virtual images of structures are provided by DT throughout their existence [13].

The idea of a digital twin is not entirely new. Its beginnings can be traced back to 1970. NASA was the first to create the term "twin". NASA created two identical spacecraft so that the conditions on Earth could be mirrored, simulated, and predicted by the conditions in space. The vehicle that remained on Earth was the identical twin of the spacecraft [8].

The DT was first born in the aerospace field and only recently has been adopted also in manufacturing contexts: such a term is used in industrial environments and in governmental research initiatives; however, scientific literature that describes the contextualization of the concept in the manufacturing domain is still at its infancy [13].

Michael Grieves, a product designer, coined the term in the early 2000s, and it was originally founded in production engineering. However, since its origin, the notion has widened and loosened to the point where it is currently being utilized, or rather exploited, to describe a variety of digital simulation models that operate alongside real-time processes in both social and economic systems also physical systems [14].

To model the behavior of engine parts at different speeds, a digital twin of an engine block was constructed in 2016. The creation of the digital twin has accelerated in recent years.

Maserati employs digital twins to cut the cost of developing new vehicles.

Furthermore, digital twin technology is being employed at ports to develop networked control technologies. Gartner predicts that 50% of the world's major industrial enterprises will use digital twins effectively by 2020, owing to the ever-increasing number of sensors. This scenario will only get worse in the future [15].

In several applications connected to future power systems, the digital twin plays an important role. It promises to be an accurate, dependable, and quick decision-making solution for stakeholders at all levels [16].

The European Center for Medium-Range Weather Forecasts (ECMWF) published an article on March 1, 2021. In this article, it is envisaged to make a digital twin of our planet. On March 5, 2021, the European Weather Forecast Center presented the concept of a digital world during a symposium. The initiative is scheduled to begin in 2021 and endure for seven to ten years. This concept was proposed in order to tackle climate change. It is planned to be used, for example, to assess the possibility of extreme droughts in various regions and how they may be mitigated. By December 2023, the first two digital twins should be available. One of them will be an extreme digital twin that is both airborne and geophysical. Environmental extremes will be assessed and estimated using this digital twin's capabilities and services. A digital twin will be created for climate change adaption. This will aid in the testing of predicted scenarios in support of regional and national climate adaption programs. Its goal is to create a real-time, highly detailed digital duplicate of the globe that includes the consequences of human activity. Observations will be gathered from a variety of sources. Individual digital twins will be more detailed than anything before seen. This will enable us to glimpse into the past, comprehend the present, and forecast the future [17].

4 Digital Twin in Industry 4.0

The rapid development of manufacturing, combined with rising demand as the economy grew, created challenges in the availability of raw materials and materials, as well as mass production applications, at the end of the 1960s. The Material Requirements Planning (MRP) system was also utilized for the first time around these dates, as a result of the widespread use of computers in businesses. By digitizing the part information of manufactured items, as well as the data on the raw material and material supply necessary for these parts, inventory holding costs and lead times have been lowered. The factory capacity was eventually added to the module, however during this time, due to the effects of globalization, the demand was defined by the client, resulting in production based on demand. While this condition results in smaller lot sizes and shorter cycle times while increasing product variety, the production structure becomes more complex. On the other hand, it necessitates the use of marketing and sales data prediction in the production equation. When MRP failed to provide this equivalence in the 1980s, the Production Resource Planning technique was developed [18]. Production Resources Planning enabled the management of data from production, planning, sales, finance, stock, marketing, logistics, and engineering activities through a single database.

Changes in the market have accompanied these advancements, resulting in the abandoning of traditional production methods and the introduction of new requirements. With the concepts of Computer Integrated Business, Computer Integrated Manufacturing, Automation, and Just in Time, such competition has enabled the technical and organizational restructuring of production systems, which reduces production costs, shortens the production cycle, and improves product quality [19]. Computer Integrated Manufacturing (CIM) systems, which integrate the product development and manufacturing processes, and Distribution Resource Planning (DRP) systems, which allow businesses to manage product distribution channels, have emerged as a result of this process.

All technological advancements have decreased the distance and time needed for worldwide communication, allowing firms to have facilities, suppliers, and consumers in multiple countries. The development of the Enterprise Resource Planning (ERP) system, which offers the integration of all ongoing information flow in an institution [20], has led to the development of all processes of production and service companies in an integrated manner, as the number of facilities, suppliers, and customers in different geographies pressured companies to organize information simultaneously.

Fig. 2 Timeline

With the integration of Customer Relationship Management (CRM), Supply Chain Management (SCM), and Business Intelligence (BI) ideas, ERP systems coupled with the Internet and call centers have evolved into the ERP II concept.

Despite the usage of a variety of specialized software programs such as MES or ERP systems, every breakdown in a subsystem reduces the whole system's efficiency significantly. Even the tiniest discrepancy between real and stored digital data causes discrepancies in planning and underestimation of the efficient point. With the paradigm shift in production and management processes brought about by the notion of Industry 4.0, it is now possible to gather real-time data, process it, and feed it into production and management decision-making mechanisms through transforming it into meaningful data. The digital twin is a critical technology for Industry 4.0 in this context.

It assesses all possible scenarios in the physical object and permits the product to make its own decisions. Industry 4.0, with its digital twin technology, revolutionizes the manufacturing process, allowing smart factories of the future to be built at a cheaper cost [21] (Fig. 2).

Following a survey of the industrial industry, Tao [22] accepted the fact that an uniform Digital Twin architecture is critically needed. A reference model is a simplified abstract architecture that provides components in schematic form. Tao et al. [12] proposed a Digital Twin reference model that included three key components: Communication, Physical Objects, and An information model for representing physical requirements and a data processing module for constructing a live depiction of a physical thing to make up a Digital Twin. Helu et al. [23] proposed a reference architecture for integrating heterogeneous manufacturing systems for the digital thread, including the capacity to diagnose root problems and govern design and manufacturing in the digital domain. Bevilacqua et al. [24] suggested a Digital Twin reference model for risk reduction in process plants by keeping users out of physical space. Shao and Helu from the National Institute of Standards and Technology [25] recently presented an overview of the Digital Twin concept for individualization in the industrial business to develop fit-for-purpose.

Since early simulation was limited to particular applications, the Digital Twin is the next wave in modeling and simulation, as seen in Fig. 3. Following that, standard tools for certain design and engineering themes were introduced in the simulation wave. Before the coming Digital Twin idea in the Industry 4.0 era [26], multilevel simulation systems were the most recent revolution. A Digital Twin can also be

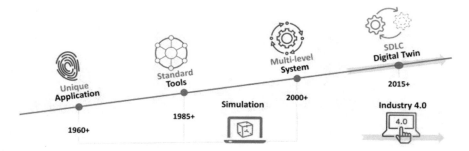

Fig. 3 Digital twin evolution

created by using a simulation and model component to replicate real-world pictures in cyberspace [27].

The German Electrical and Electronic Manufacturers' Association established the Reference Architecture Model for Industry 4.0 (RAMI 4.0) to support Industry 4.0 activities based on a holistic view of manufacturing firms, given its growing importance and effect in the Industry 4.0 age. RAMI 4.0 [28] provides enterprises with a complete framework for building future products and business models utilizing a three-dimensional map in a systematic manner.

The underlying structure of elements, relationships, interfaces, processes, limitations, principles, purposes, physical, and logical properties are all addressed by architecture [29].

In Industry 4.0, Fig. 4 depicts a Digital Twin reference architectural model. This three-dimensional layered model is made up of three-dimensional coordinates that describe all important aspects of the Digital Twin. Complex interrelations, such as the five Digital Twin layers, agile value life cycle [30], and Digital Twin integration hierarchy [31], are broken down into smaller and simpler clusters in this way.

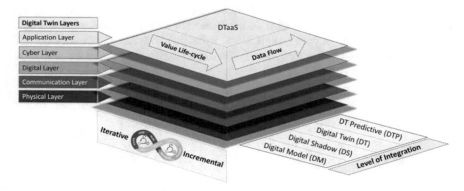

Fig. 4 Digital twin reference architectural model

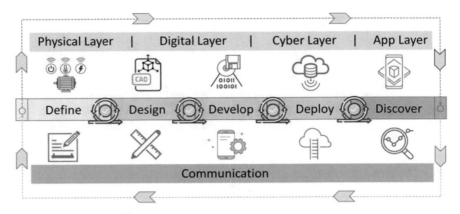

Fig. 5 Left axis in digital twin reference architectural model

This section looks at the left axis in Fig. 5. The five Digital Twin levels are represented by the vertical axis, which are made up of an interconnected set of physical, communication, digital, cyber, and application layers.

The Digital Twin layers in the proposed reference architectural model must coincide with the 5D product design process, as shown in Fig. 6. "Define, Design, Develop, Discover, and Deploy" are the five "D"s. A remedy for a physical/real problem, need, or wish will be done in the "Define" phase. "Design" and "Development" are participative phases in the Digital layer that result in CAD and CAM file formats to fulfill real-world expectations for digital versions of products, services, and processes. We may use Big Data to collect data from the physical layer and begin giving personalization. As a result, through an iterative, incremental process, cyberspace enables rolling products and services in terms of modifications, revisions, upgrades, predictions, UX/CX design, and individualization.

This section looks at the right axis in Fig. 6. The four layers on the right axis represent the various levels of Digital Twin integration. Although the phrases Digital Model, Digital Shadow, and Digital Twin are sometimes used interchangeably, data integration toward the physical, digital, and cyber layers is not the same. Figure 6 depicts a "Digital Model," which, unlike simulation and mathematical models, does

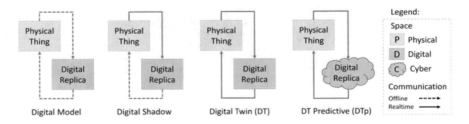

Fig. 6 Right axis in digital twin reference architectural model

not have a real-time connectivity to and from a physical object. If a change in the state of a thing causes an instantaneous change in the state of digital things, then one-way real-time data exchange from physical to digital space creates a "Digital Shadow," a useful tool in real-time monitoring.

A twin is created by bidirectional real-time data transmission between physical and digital space. Digital twin enabled implement in manufacturing planning and control, maintenance, and layout planning develop as a "Digital Twin" integration. A Digital Twin Predictive' (DTp) is a digital clone of a physical object that communicates in two dimensions in real-time over the internet [32].

5 Digital Twin Applications

Digital twin integrates the virtual models in virtual space and physical objects in physical space in manufacturing and provide a promising opportunity to implement smart manufacturing and industry 4.0. Utilization of digital twins in manufacturing is also seen as an opportunity to achieve a higher level of productivity. In this perspective, use of autonomous systems to control and manage the available data and create a copy of real manufacturing processes in a real-time digital environment is provided by digital twin. This means the results of the manufacturing process can be seen without the need of physical systems or environment. Digital twins can be used in every step of the manufacturing cycle that includes logistics, development of the product, creating a baseline system and production [33].

In plant logistics, the use of real-time data is an important instrument for visualizing events immediately [34]. Logistic firms rely on digital twins to track and analyze; key performance indicators, such as packaging performance, fleet management, and route efficiency. Furthermore, by industry 4.0 factories will cope with more complex and variable material and handling problems. Digital twin can be used to integrate advanced storage and handling systems via AR technologies. For example, DHL uses wearable devices such as Google Glass Enterprise Edition or Microsoft HoloLens as virtual reality training tools for picking systems [35].

In product development phase, digital twin of products enable better understanding of requirements and specifications of future products. Data from the digital twin of previous products can be used to refine the requirements and specifications of future ones. Physical tests can be conducted with digital twins and simulation of the product behavior can be detected with lower development cost with higher reliability.

Digital twin can be also used at maintenance to identify the impact of state changes on upstream and downstream processes of a production system, to evaluate of machine conditions based on descriptive methods and machine learning algorithms and integrate, manage and analyze machinery or process data during different stages of machine life cycle to handle data/information more efficiently and further achieve better transparency of a machine's health condition [36]. Aivaliotis et al. [37] uses the digital twin as a predictive maintenance tool. In a production plant system has been simulated before inputting the data gathered from machines via smart control system. Product failure analysis and prediction, and product maintenance strategy that have in common the analysis of the real-time state data and historical data to predict a fault and construct a maintenance strategy. The result for the simulation is used to predict the health of the machine and predict the maintenance activities. With this method conditions of the machines can be understood without disturbing machines' work.

CNH Industrial has used digital twin to optimize maintenance at its plant in Suzzara, Italy, where it produces Iveco vans. The problem was the insufficient reliability of welding machines on the chassis line. The goal was to improve reliability by utilizing digital twin of the line. Component failure probability is estimated by simulation and machine learning. The main objective is to optimize maintenance costs while minimizing the idle time due to spare parts and maintenance [38].

Another example is Baker Huges, an oil and gas equipment producer. They are also utilized digital twin to improve performance of the company as well as reacting to the nonconforming issues immediately [39].

Digital twin systems can also be used to identify faults in the system. Wang et al. [40] developed a digital twin application for rotating machinery fault identification. For this purpose the finite element model of the rotor is constructed via using geometry, physical properties and the physics of the system. In this application critical speed and unbalance response investigated under different conditions. With digital twin accurate diagnosis can be made compared to the traditional methods.

Lastly, digital twin can also be used for layout optimization. A case study in a welding production workshop is studied. Digital twin enable situation analysis, optimization plan making, virtual simulation, optimization plan implementation, and result analysis for layout optimization [41].

In conclusion, manufacturing operations have been a particular area of focus for digital twin development and digital twin can be utilized in every phase of manufacturing from logistics to prototype. Even though digital twin is not yet utilized fully, it is a very promising tool for optimization and productivity.

6 Digital Twin Challenges

This chapter describes the challenges of digital twins.

- It Infrastructure

Similarly, the current IT infrastructure is a problem for both analytics and IoT. The Digital Twin requires infrastructure that enables IoT and data analytics to succeed; these will aid in the efficient operation of a Digital Twin. The Digital Twin will not be able to achieve its objectives without a well-connected and well-thought-out IT infrastructure.

- Useful Data

The data required for a Digital Twin is the next challenge. It must be high-quality data with no noise and a continuous, uninterrupted data stream. If the data is bad and inconsistent, the Digital Twin may underperform since it is acting on bad and missing data. For Digital Twin data, the quantity and quality of IoT signals are critical. Planning and analysis of device usage are required to determine which data should be collected and used to make the most of a Digital Twin.

- Privacy and Security

It is apparent that the privacy and security issues related with Digital Twins are a problem in the workplace. First, because of the large amount of data they use, and second, because of the potential risk to critical system data. To address this issue, the primary enabling technologies for Digital Twins–data analytics and IoT–must adhere to current security and privacy policies and legislation. Consideration of security and privacy for Digital Twins data aids in the resolution of trust concerns with Digital Twins.

- Trust

The issues connected with trust can be seen from both an organizational and a user's perspective. To ensure that end-users and organizations understand the benefits of a Digital Twin, which will strive to solve the barrier of trust, the technology must be addressed further and presented at a fundamental level. Another method for over-coming trust issues is model validation. It is critical to ensure user trust by ensuring that Digital Twins perform as expected. Trust in Digital Twins grows as more is learned about them. The enabling technology will provide more visibility into the actions used to ensure that privacy and security practices are followed throughout development, hence addressing trust issues.

- Expectations

Despite industry giants Siemens and GE accelerating Digital Twin deployment, caution is essential to highlight the hurdles that exist for Digital Twin aspirations and the need for further understanding. The requirement for robust foundations for IoT infrastructure, as well as a better knowledge of data required for analytics, will ensure that businesses embrace Digital Twin technologies. It is also difficult to fight the notion that the Digital Twin should only be employed because of current trends. The benefits and drawbacks of Digital Twin expectations must be explored in order to take proper action when establishing Digital Twin systems. It's apparent that the issues faced by Industrial IoT/IoT and data analytics are likewise challenges faced by the application of a Digital Twin. Despite the issues that Digital Twin shares with IoT and data analytics in terms of user experience, privacy, and infrastructure, there are also unique challenges related to the modeling and construction of the Digital Twin.

- Standardized Modeling

Because there is no common approach to modeling, the next issue in all forms of Digital Twin development is modeling such systems. There must be a uniform approach from the initial concept to the simulation of a Digital Twin, whether it is physics-based or designed-based. Standardized procedures ensure domain and user understanding, as well as information flow between stages of a Digital Twin's development and implementation.

- Domain Modeling

Another issue arising from the necessity for uniform use is ensuring that information about domain use is communicated to each of the development and functional stages of a Digital Twin's modeling. This enables compatibility with sectors like IoT and data analytics, allowing the Digital Twin to be used successfully in the future. These are critical moving ahead since they ensure that they be taken into account in the establishment of Digital Twins as well as when using IIoT/IoT and data analytics.

6.1   Benefits of Digital Twin

The advantages of the digital twin are discussed in this section. The advantages of using a digital twin vary based on when and where you utilize it. Using the digital twin to monitor existing products like a wind turbine or an oil pipeline, for example, can minimize maintenance costs and save millions of dollars. Digital twins can also be used to prototype before going into production, reducing product errors and speeding up the time to market. Process improvements, such as monitoring workforce levels by output or matching a supply chain with production or maintenance requirements, are some examples of how a digital twin might be used.

Increased reliability and availability can be achieved by using monitoring and simulation to improve performance. They can also reduce the risk of accidents and unscheduled downtime due to failure, lower maintenance costs by forecasting failure before it happens, and ensure that production targets are not harmed by maintenance, repair, and replacement parts orders. By evaluating customization trends and ensuring product quality through real-time performance testing, the digital twin can also bring continual upgrades.

However, despite its many advantages, the digital twin is not appropriate in all scenarios since it might add to the level of complexity. Some business challenges may not necessitate the use of a digital twin and can be solved with minimal effort and expense.

References

1. Stargel EG (2012) The digital twin paradigm for future NASA and U.S. Air force vehicles. içinde
2. Chen Y (2017) Integrated and intelligent manufacturing: perspectives and enablers (s. pp 588–595). içinde
3. Mrad ZL (2018) The role of data fusion in predictive maintenance using digital twin
4. Intell YZ (2018) An application framework of digital twin and its case study (s. pp 1141–1153). içinde
5. Roy RV (2018) Digital twins: understanding the added value of integrated models for through-life engineering services
6. Lucero AM (2019) Leveraging digital twin technology in model-based systems engineering
7. Sihn WK (2017) Digital twin in manufacturing: a categorical literature review and classification
8. Barricelli BR, Casiraghi E, Gliozzo J, Petrini A, Valtolina S (2020) Human digital twin for fitness management. IEEE Access 8:26637–26664. https://moh-it.pure.elsevier.com/en/public ations/human-digital-twin-for-fitness-management. adresinden alındı
9. Schmich M (2016) How siemens PLM is prepared to help customers' digital transformation I thought leadership. January 7, 2022 tarihinde Siemens Blogs: https://blogs.sw.siemens.com/ thought-leadership/2016/06/21/how-siemens-plm-is-prepared-to-help-customers-digital-tra nsformation/ adresinden alındı
10. Digital Twin for the Digital Power Plant (2016) 01 07, 2022 tarihinde GE Digital Twin: Analytic Engine for the Digital Power Plant: https://www.ge.com/digital/sites/default/files/download_ assets/Digital-Twin-for-the-digital-power-plant-.pdf adresinden alındı
11. Liu M, Fang S, Dong H, Xu C (2021) Review of digital twin about concepts, technologies, and industrial applications. J Manuf Syst 58:346–361. https://www.sciencedirect.com/science/art icle/pii/S0278612520301072 adresinden alındı
12. Tao F, Sui F, Liu A, Qi Q, Zhang M, Song B, Guo Z, Lu SC, Nee AY (2018) Digital twin-driven product design framework. Int J Prod Res 57(12):3935–3953. https://doi.org/10.1080/ 00207543.2018.1443229
13. Negri E, Fumagalli L, Macchi M (2017) A review of the roles of digital twin in CPS-based production systems. Procedia Manuf 11:939–948. https://www.sciencedirect.com/science/art icle/pii/S2351978917304067#aep-article-footnote-id3 adresinden alındı
14. Batty M (2018) Digital Twins. Environ Plann B Urban Anal City Sci 45(5):817–820. https://journals.sagepub.com/doi/full/https://doi.org/10.1177/2399808318796416?_cf_chl_ jschl_tk__=aJjZw_hrcaqXJh1NG8eQBws0Rz1Jhdz_0iMIoeqF8IA-1641576345-0-gaNycG zNEKU. adresinden alındı
15. Uttendorfer U, Hornberg O (tarih yok) (2022) Consulting: digital twin—UNITY. January 7, 2022 tarihinde UNITY consulting and innovation: https://www.unity.de/en/services/digital-twin/ adresinden alındı
16. Palensky P, Cvetkovic M, Gusain D, Joseph A (2021) Digital twins and their use in future power systems [version 1; peer review: 1 approved, 1 approved with reservations]. Digit Twin 1(4). https://digitaltwin1.org/articles/1-4/v1 adresinden alındı
17. Presenting Destination Earth: A Digital Replica of Our Planet (2021). January 7, 2022 tarihinde ECMWF: https://www.ecmwf.int/en/about/media-centre/news/2021/presenting-des tination-earth-digital-replica-our-planet-adresinden-alındı
18. Xiao C (2017) The reason and coping measures of employees' resistance to information system. J Hum Resour Sustain Stud 5(1):141–176
19. Hozdić E (2015) Smart factory for industry 4.0: a review. Int J Mod Manuf Technol 7(1):28–35
20. Rajagopal P (2002) An innovation—diffusion view of implementation of enterprise resource planning (ERP) systems and development of a research model. Inf Manage 40(2):87–114
21. Kumaş E, Serpil E (2021) Endüstri 4.0'da Anahtar Teknoloji Olarak Dijital İkizler. J Polytech 24(2):691–701
22. Tao F (2018) Digital Twin in industry: state-of-the-art. IEEE Trans Industr Inf 15(4):2405–2415

23. Helu M (2017) Reference architecture to integrate heterogeneous manufacturing systems for the digital thread. CIRP J Manuf Sci Technol 19:191–195
24. Bevilacqua, M. (2020). Digital Twin reference model development to prevent operators' risk in process plants. Smart Prod Oper Manage Indus 4.0 12(3):1088
25. Shao G (2020) Framework for a digital twin in manufacturing: scope and requirements. Manuf Lett 24:105–107
26. Rosen R, Wichert GV (2015) About the importance of autonomy and digital twins for the future of manufacturing. IFAC-PapersOnLine 48(3):567–572
27. Ghosh AK (2019) Hidden Markov model-based digital twin construction for futuristic manufacturing systems. AI EDAM 33(3):317–331
28. Pisching MA (2018) An architecture based on RAMI 4.0 to discover equipment to process operations required by products. Comput Ind Eng 125:574–591
29. Pressman R (2009) Software engineering a practitioner's approach (7 b)
30. Aheleroff S (2019) The degree of mass personalisation under industry 4.0. Procedia CIRP 81:1394–1399
31. Kritzinger W (2018) Digital Twin in manufacturing: A categorical literature review and classification. IFAC-PapersOnLine 51(11):1016–1022
32. Aheleroff S, Xu X (2021) Digital twin as a service (DTaaS) in industry 4.0: an architecture reference model. Adv Eng Inf 47:101225
33. Cimino C, Negri E, Fumagalli L (2019) Review of digital twin applications in manufacturing. Comput Indus 113. https://doi.org/10.1016/j.compind.2019.103130 adresinden alındı
34. Park H, Srinivasan A, Bolsius R, Hsiao E, Toillion J, Shukla V (tarih yok) Real-time monitoring event analysis and monitoring
35. DHL Supply Chain deploys latest version of smart glasses worldwide. (2019). January 7, 2022 tarihinde Deutsche Post DHL Group: https://www.dpdhl.com/content/dam/dpdhl/en/media-rel ations/press-releases/2019/pr-smart-glasses-20190521.pdf adresinden alındı
36. Kritzinger W, Karner M, Traar G, Henjes J, Sihn W (2018) Digital twin in manufacturing: a categorical literature review and classification. IFAC 51:1016–1022
37. Aivaliotis P, Georgoulias K, Chryssolouris G (2017) A calculation approach based on physical-based simulation models for predictive maintenance. In: International conference on engineering technology and innovation
38. Digital Twin of a Manufacturing Line: Helping Maintenance Decision-making. (tarih yok). anylogic: https://www.anylogic.com/digital-twin-of-a-manufacturing-line-helping-mai ntenance-decision-making adresinden alındı
39. Baron K (tarih yok) Seeing double: digital twins make GE and baker hughes supply chain inno-vators | *GE News*. January 7, 2022 tarihinde General Electric: https://www.ge.com/news/rep orts/seeing-double-digital-twins-make-ge-baker-hughes-supply-chain-innovators adresinden alındı
40. Wang J, Ye L, Gao RX, Li C, Zhang L (2018) Digital twin for rotating machinery fault diagnosis in smart manufacturing. Int J Prod Res
41. Gou H, Zhu Y, Zhang Y (2021) A digital twin-based layout optimization method for discrete manufacturing workshop. Int J Adv Manuf Technol 112:1307–1318. 10.1007/s00170-020-06568-0adresindenalındı

Big Data Analytics in Industry 4.0

Mustafa Bugra Ozcan, Batıhan Konuk, and Yıldız Merve Yesilcimen

Abstract With the unpredictable development of technology, a wide variety of data is produced in a very short time from countless sources. Industry 4.0 is a revolution in new technology for the digital world of digital factories and smart products. The 'Big Data' used by machines to communicate with each other using the internet of things is the building block of Industry 4.0. Big data is not just data to be stored or accessed. It is also data that is made sense and analyzed. In this chapter, the characteristics of big data, its applications, and its place in industry 4.0 are presented. Firstly, history and definitions of the big data are introduced in this chapter, then some applications are explained by using big data methods and techniques. Finally, opportunities and challenges are mentioned for the future aspects of using big data analytics.

1 Introduction

193187495 bits of data is created by the time you finish reading this sentence, assuming you really tried to read the exact number. 5.7 million searches, 315 thousands photos, and 6 million people shop online while you came up to the end of the second sentence [32]. These are just some 'little' information to keep in mind while reading this chapter. Data can be defined as collection of raw information particles such as facts or numbers that can be processed, transmitted, and stored to help decision-making, gain experience, or do analyses. It can be structured or unstructured, machine readable or not, digital or analogue, and personal or not. Over the last years, the unprecedented increase of computing capabilities and generation of new data give rise to birth of the term big data which cannot handle classical approaches, methods, or tools.

Data analytics was first utilized in Mesopotamia over 7000 years ago to monitor and regulate the growth of crops and livestock, according to the oldest documents. With more applications and contributions to human history, this idea has grown

M. B. Ozcan · B. Konuk · Y. M. Yesilcimen (✉)
Department of Industrial Engineering, Hacettepe University, Beytepe Campus 06800, Ankara, Turkey
e-mail: m.yslcmn@gmail.com

© The Author(s), under exclusive license to Springer Nature Singapore Pte Ltd. 2023   171
A. Azizi and R. V. Barenji (eds.), *Industry 4.0*, Emerging Trends in Mechatronics,
https://doi.org/10.1007/978-981-19-2012-7_8

throughout the years [69]. The method is defined by big data analytics, which collects, organizes, and analyzes enormous datasets to discover patterns and other important information. Ordinary data is different, more complicated, and more large-scale new to discover significant hidden values from datasets using big data analytics. It is a collection of technologies and methods that necessitate some level of integration [89].

Industry 4.0 refers to a digital revolution that includes digital factories and smart gadgets. It is viewed as a new industrial phase in which many developing technologies are merged to give digital solutions. The Internet of Things (IoT) has brought the world of machines closer together by allowing objects to communicate with one another and access the Internet. The Internet of Things now connects all of the machinery in an Industry 4.0 factory. In these systems, big data analytics opens up new opportunities [37, 55, 76].

1.1 History of Big Data

In his book 'The Industries of the Future,' Alec Ross says 'Land was the raw material of the agricultural age. Iron was the raw material of the industrial age. Data is the raw material of the information age' [73].

Big data is drastically growing all around the world [41]. As it is known, data has played a significant role in human life. End users of data generate data using á variety of devices, which record an ever-increasing number of occurrences [38]. It is also continually generated from social media streams, search engines, digital images, banking and dealing records, sensors, GPS signals, and uncounted different sources [41].

Big data comes with big history. There are different opinions in different sources about the birth of big data [12]. According to Diebold et al. [30], big data was first mentioned by computer scientist John Mashey at Silicon Graphics Inc. in the mid-1900s. However, no academic study has yet been published at that time. The first notable academic studies, independently of each other and Silicon Graphics, were by [93], in computer science in 1998 and by Diebold [29], in statistics/economics in 2000. Diebold states unpublished and non-academic research note. Douglas Laney in 2001 also contributed significantly to this subject. Consequently, Diebold attributes the term 'big data' to Mashey, Weiss, Indurkhya, Diebold, and Laney [30].

Another view regarding the emergence of the term 'big data' was put forward by Wang et al. [92]. According to them, the term 'big data' was first used by Michael Cox and David Ellsworth to describe data visualization and the problems it brings to computer systems in a 1997 paper presented at an IEEE conference [21].

In the late 1990s, rapid IT innovations and technology improvements generated large volumes of data, but little usable information was available. The years 2001–2008 were critical in the development of big data. Big data analytics reached the breakthrough period at the beginning of 2009. The utilization of cloud in association

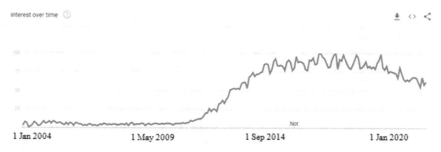

Fig. 1 According to Google trends data interest in 'big data' from 2004 to present [14]

with data has become a more modern trend in big data analytics technologies. Companies are increasingly adopting 'big data in the cloud' solutions like software-as-a-service (SaaS), which provide a cost-effective option. The amount of data produced by sensors will continue to increase considerably [21] (Fig. 1).

Although the history of big data dates back to the mid-1990s according to references, when Google Trends 'big data' searches are taken into account, as seen in Fig. 1, it is seen that the term became widespread after 2010 [14]. The interesting point is that Google Trends uses big data to provide us with information about how the interest in big data changes over time [52].

1.1.1 Evolution of Big Data and Data Analytics

Large-scale data collection and storage extend back to the early 1950s, when the first industrial computers were launched. Due to the expense of computers, storage, and data networks from the early 1950s through the mid-1990s, data expansion was not very high. The data is heavily organized throughout this time period, mostly to support operational and transactional information systems. The early 1990s saw the birth of the World Wide Web (www), which facilitated massive data expansion and the growth of big data analytics. Big data and data analytics have gone through three stages since the invention of the World Wide Web: Big Data 1.0, Big Data 2.0, and Big Data 3.0 [58].

- Big Data 1.0 (1994–2004)
  In 1994, Big Data 1.0 was released, coinciding with the birth of e-commerce. Online businesses were the primary suppliers of web content throughout those years. Because of the technological limitations of online applications, user-created material constituted just a small percentage of the overall web content. Web mining techniques have been expanded in this period to evaluate people's online behavior.
- Big Data 2.0 (2005–2014)

The Web 2.0 and social media phenomena are accelerating growth. Introducing Big Data 2.0. Web 2.0 is an online paradigm that evolved in the 1990s as a result of web technology, allowing web users to connect to websites and provide their own content. Social media analytics enables content mining, use mining, and structure mining. Social media analytics evaluate and analyze human behavior on social media platforms, providing consumers with insights and conclusions based on their interests, Internet browsing behaviors, friend networks, moods, occupation, and perspectives. By gaining a deeper understanding of their consumers through social media analytics, businesses can focus their relationship marketing efforts on specific consumer groups and tailor products and services to their individual requirements and interests.

- Big Data 3.0 (2015–)
  Big Data 3.0 combines Big Data 1.0 and 2.0. The key contributors to Big Data 3.0 are IoT apps that generate data in the form of images, audio, and videos. The Internet of Things (IoT) is a technical ecosystem in which items and sensors are assigned unique identifiers and are able to exchange data and collaborate over the Internet without requiring human contact. Due to the IoT's rapid growth, connected devices and sensors will soon supplant social media and e-commerce websites as the primary producers of big data [58].

1.2 What Is Big Data?

Big data, in the existing literature, is the 'next big thing in innovation' [6].

According to October 2021 data, 4.88 billion people worldwide use the Internet, which makes up about 62% of the global population. This number is also increasing. However, because the coronavirus epidemic has had a significant influence on Internet user research, actual growth figures could be substantially higher than this rate implies [24].

McKinsey Global Institute defined that the big data is datasets that capture, manage, store, process, and analyze data with low latency; the size and type of which are beyond traditional relational databases [61].

Microsoft, on the other hand, defined it as data that strains the capacity of the traditional data management system due to its very large size, too much complexity and high speed [75].

Big data was declared the next frontier for productivity, innovation, and competition in May 2011 [61]. Due to the rapid development and evolution of big data literature, there is no universal and precise definition of concept of big data [66]. Different researchers have different opinions on data characteristics. Therefore, in this chapter, definitions of big data are given. When data becomes too big for traditional systems to manage, it is referred to as 'big data.' It is not only about size when it comes to being big [6].

1.3 Application Areas of Big Data

Companies can make better business or increase profit with the help of applications of big data. Analyzing collected data, discovering new patterns, having insights of trends, market analyses, predicting customer choices, etc., can be several advantages for companies. In today's world, taking advantage of these benefits is not as difficult as seen, because there are so many as well as great sources of data. Social media, capture of sensors, or customer reviews are just a few big data sources for organizations. Several big data applications that can adapt these sources can be stated as follows [85].

- Health care,
- Basic sciences,
- Media and entertainment,
- Government applications,
- IoT,
- Manufacturing.

The increase in wearable health products with mobile applications and digital databases for health industry give ability to use big data analytics for health care such as drag side-effects, treatments, and other information related to patient analyses [31]. Electronic health records instead of paper-based archives play an important role for this start [85]. Marx mentioned its applications to biology such as genetic sequences, interactions of proteins, or findings in medical records with its own challenges such that heterogeneity in biological data, requiring prior knowledge to interpret, or storing problems since even a single sequenced human genome is around 140 GB in size [65]. In [26], they emphasized the importance of flows rather than stocks in data science by showing continuous flows are the game changer in comparison with traditional methods. They also divided its applications into three categories which are customer-facing processes, monitoring the processes, and exploring relationships.

Another paradigm shift can be seen in the manufacturing industry. From classical manufacturing to automated manufacturing is now furthered to smart manufacturing. During this evolution, data has an important factor on processes [23] (Fig. 2).

In this chapter, we mainly focus big data analytics and its applications to Industry 4.0.

2 Big Data Characteristics

The term 'big data' was commonly used to describe the amount of data that standard database methods and tools could not handle efficiently. The concept of 'V' is used by some academics to define 'big data' [53]. Big data is defined as the ability to manage large amounts of different data at the exact speed and in the best timing

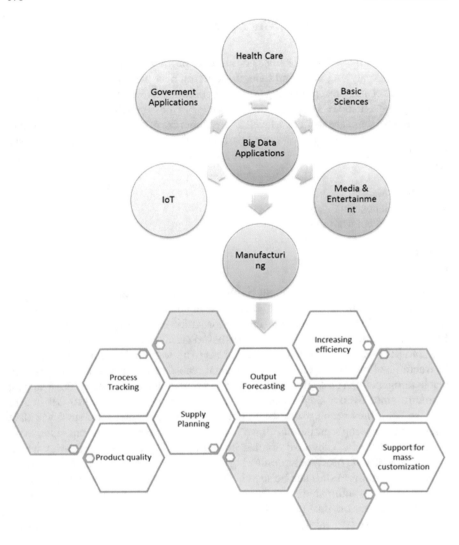

Fig. 2 Big data application areas and its expansion on manufacturing industry

to allow real-time analysis and reaction. The three Vs have become a typical way of describing big data. Thus, it was first defined in terms of volume, velocity, and variety (3Vs) [13, 18, 39, 46, 57] (Fig. 3).

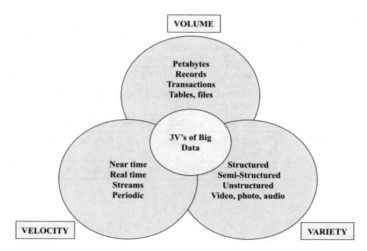

Fig. 3 The 3V's of big data

2.1 *Volume*

When pondering the question 'what is big data?', size is undoubtedly the first attribute that comes to mind. Volume refers to the enormity of data. There is not much consensus on how big data must be in order to be considered 'big data' [38].

Computers use data in 0 or a 1, each of which is called a 'bit.' A 'byte' consists of eight bits, and this is a single unit of storage in a computer's memory. A single number, letter, or symbol is represented by a byte. Different concepts are used to name large amounts of data such as [7]

- Gigabyte (1.0×10^9 bytes)
- Terabyte (1.0×10^{12} bytes)
- Petabyte (1.0×10^{15} bytes)
- Exabyte (1.0×10^{18} bytes)
- Zettabyte (1.0×10^{21} bytes)
- Yottabyte (1.0×10^{24} bytes).

Exabytes and zettabytes are the units of measurement for the quantity of data that will be generated in the future [27] (Table 1).

Table 1 Data storage size examples

Unit	Dimension	Example
Gigabyte	138.402 GB	Internet traffic in one second [79]
Terabyte	74 TB	The amount of data in the US Library of Congress in 2009 [83]
Petabyte	2.5 PB	The memory capacity of the human brain, according to scientists' estimates [83]
Exabyte	56.8 EB	Global monthly mobile data traffic in 2021 [20]
Zettabyte	44 ZB	The amount of data in the world at the start of 2020 is forecasted [45]
Yottabyte	1 YB	257.054 trillion DVDs—a basic DVD is roughly 4.7 GB in size [83]

By 2025, forecasts suggest the following:

- The Internet is expected to surpass the brain capacity from everyone on the earth [27].
- According to IDC (International Data Corporation) estimates, the amount of data in the global data sphere is predicted to be 175 zettabytes [72].
- More than 463 exabytes of data will be created worldwide per day, the equivalent of 212,765,957 DVDs per day [28].
- According to Statista, the number of Internet of Things (IoT) devices on the world will reach 75 billion [82].

2.2 Velocity

In current history, as the quantity of data has increased, the rate at which it becomes accessible has also increased considerably [7]. The velocity dimension of big data is used to indicate the frequency or speed of data produced or delivered [74]. Data velocity is measured by the rate at which data is created, streamed, and collected [33].

Industry 4.0 is based on the use of digital technology to collect and analyze data in real time, delivering important data to the manufacturing system [37]. Therefore, big data enables data generation, analysis, and response in real time or near real time [90].

Even though data storage capacity might be increased, the rate at which new data is generated is more crucial to consider. Even if data is available, business possibilities may be lost until it can be processed in real time. For example, if weather predictions are delayed due to a slower processing speed that cannot keep up with the rate at which data is received, it has an impact on making the right judgments at the right time [88].

There are three different types of big data applications: online service, offline analytic, and real-time analytic. Data velocity refers to the speed with which data is processed in online services such as video streaming. On the other hand, data velocity refers to the frequency with which data is updated in offline analytic (such as Bayes classification or sort) and real-time analytic (such as relational database queries like select and join). All of the above-mentioned data velocities should be controlled by big data producers [67, 96]. According to Anderson [7], fast-received data can be classified in the following two ways:

- Streaming data
- Complex event processing.

The challenge of velocity comes from the need to cope with the speed of newly created or updated existing data [22]. A number of data integration strategies have been developed primarily to handle static data. Unfortunately, these methods are inefficient at handling and integrating IoT streaming datasets from various sources [87].

Streaming data is data that is sent to an application at a very fast rate. For example, content downloaded and watched from Netflix and Amazon can be given. In these cases, the data is downloaded while playing the video. If the Internet connection is not very fast, it is likely that interruptions or glitches in the data download will occur. So more speed is needed. When real-time decisions are required, streaming comes in handy. As new market data becomes available, traders, for example, must make split-second decisions [7]. It is vital to be able to appropriately react and answer to client queries in a prompt and timely manner [87]. For example, Walmart handles over one million transactions per hour [22].

CEP (Complex Event Processing) systems are designed to process data quickly and discover interesting events as they occur [36]. An example of this is the GPS device creating new route using traffic and accident data [7].

According to Internet live stats data [1] in one second:

- 9808 Tweets sent
- 1127 Instagram photos uploaded
- 137,674 GB of Internet traffic
- 98,381 Google searches
- 93,359 YouTube videos viewed
- 3,101,985 emails sent (approximately %67 of email is spam).

2.3 Variety

The reason for variety in big data is that the data comes from a wide variety of sources including smart devices, sensors, web sources, and social media, thanks to technological advancements such as the Internet of Things (IoT) and Artificial Intelligence (AI). Therefore, there are structural differences in the data [80]. In other words, the data has structural heterogeneity. Given the diversity of data sources, heterogeneity is an inherent characteristic of big data [35, 48].

For example, unstructured data is now as frequent as structured data, and the majority of these are generated in real time, implying that complexity is now the most essential element of big data [80].

Data is often not available in a perfect and ready to process form. This data comes from a wide range of sources, resulting in massive stacks. Many companies must deal with enormous datasets containing various types of data (e.g., IoT streaming data, static data) in various formats (e.g., structured, unstructured) originating from various sources. Data analysis was started to deal with these data stacks [41, 74, 87].

Big data sources, according to Groves et al. [42] and Chen et al. [17], are categorized into five groups based on how they are created. Because the category names are nearly identical, they are explored under the same headings. These categories are as follows [17, 42]:

1. Human-generated data: Word documents, emails, spreadsheets, images, and video files are just a few examples. This data is provided directly by users and is generally not structured for user convenience. It is required to structure and extract information from raw data. This is accomplished through the use of analysis algorithms.
2. Transactional data: Web logs, business transactions, moving item feeds, sensor network reports, and RFID scans are all examples of massive transactional data. This data is usually structured according to preset standards. They are frequently gathered in a stream.
3. Scientific data: Celestial data, high-energy physics data, genetic data and finger-prints, retinal scans, blood pressure, and similar data types may also be included. Scientific data can be structured, semi-structured, and unstructured.
4. Web and social media data: Data scanned and processed to support applications such as web search and mining. Clickstream and interaction data from social media such as Twitter, Instagram, LinkedIn, and forums. It can also include health plan websites, smartphone apps, etc.
5. Graph data: The links between the nodes and the information nodes. Social networks and RDF (Resource Description Framework) knowledge bases are two examples. These are structured data.

Abawajy has done a comprehensive study on the variety of big data, and as a result of this study, he divided big data diversity into four categories. He then divided them into subcategories [2] (Fig. 4).

In addition to the 3Vs, the expanding literature identifies a number of other important features, with big data being one of them (Table 2).

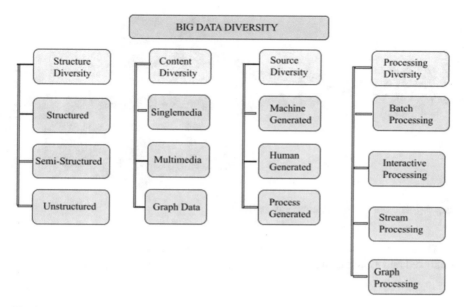

Fig. 4 Classification of big data diversity

Table 2 Big data characteristics according to different authors

Number of V's	Characteristics other than 3Vs (volume, velocity, variety)	Authors, date
4 V's	Variability	[59]
5 V's	Veracity, value	[9]
6 V's	Variability, veracity, and value	[38]
7 V's	Variability, veracity, value, and visualization	[81]
8 V's	Variability, veracity, value, validity, and volatility	[78]
11 V's	Variability, veracity, value, validity, volatility, visualization, valence, and vulnerability	[88]

2.4 Static and Streaming Data

Data can be different types such as static data and streaming data [87]. The operational difference between static and streaming data is in the storage and analysis of the data (Fig. 5).

- Static data is first stored in a database and then analyzed. After the analysis, results are reported, and the data can be stored in the database for reuse.
- Streaming data, on the other hand, is analyzed as soon as they are observed. Only the results from the analysis are reported and stored. Streaming data is discarded after analysis. Only analysis results are stored.

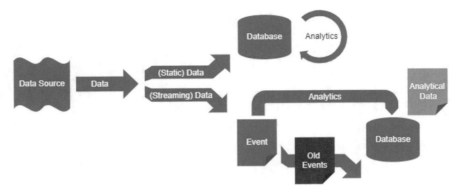

Fig. 5 Static data versus streaming data [84]

Data is both static and streaming comes from the same source. They can be analyzed and stored using same analytics and storage service [84].

2.5 Formats of Data

Not all data is created equal [47]. Organizations may now develop a range of structured, semi-structured, and unstructured data forms thanks to technological advancements [58]. If we explain these different data formats:

- Structured data: Well ordered and simply decipherable by MLAs (machine learning algorithms). It is typically classified as quantitative data [47]. They only make up 5% of all available data [22].
- Unstructured data: Data that cannot be handled or reviewed using traditional data tools and methods is referred to as qualitative data. Because unstructured data lacks a defined data model, it is best stored in non-relational (NoSQL) databases. Unstructured data may also be stored in raw form in data lakes, which is another method for handling it [47]. As new analytics technologies are developed, unstructured data is produced considerably faster than structured data, and data type becomes less of a hurdle to analysis [58].
- Semi-structured data: The 'bridge' between structured and unstructured data is semi-structured data. It is more complicated than structured data but simpler to save than unstructured data since it lacks a specified data model [47].

3   Big Data Analytics

Typical data analytics approaches can be summarized into 4 main groups as follows [85].

- Statistical modeling methods,
- Data mining schemes,
- Machine learning schemes,
- Data visualization techniques.

Statistical methods are the ones which utilize statistical theory to inference. These methods can be used for different purposes such as revealing relationships among data, generalizations from the data, or predicting future quantities. In data preprocessing steps, statistical methods can also be used for features or dimension reductions.

Data mining is a general term for methods which try to extract some patterns from massive datasets. Moreover, machine learning explores datasets and try to perform automatization.

Furthermore, data visualization is also an important approach in big data analytics. It is a collection of methods to display data in some visual forms such as charts, graphs, or maps. In order to increase readability of big data sets, these visualization techniques are useful.

Bernard has stated that the world is getting smarter [62]. In sport industry, smart technology is used to find new talents or monitor performances. Giving detailed feedback on basketball trainings, analyzing match statistics in tennis, or real-time performance assessments in rugby are a few examples for this industry. Smarter health is another area which witnesses considerable interest such as injury scanners, cancer detections, or diagnosis systems. From heating systems to gadgets, many equipment at home are also getting smarter. Dynamically alter temperature for energy-efficient purposes and prevent children to watch inappropriate content in TV are good examples for smart things at homes. Bernard even mentioned about smarter love by giving online dating site examples that match people using data analytics [62]. SMART data method is introduced after these examples in [62] which gives comprehensive description of its steps such that

Start with strategy,
Measure metrics and data,
Analyze your data,
Report your results,
Transform your business and decision-making.

Start with strategy is the first step for handling big data and it recommends determining the aim to use data regardless of size. Second step is discovering the data such as its types, its sources, and our needs for it with appropriate metrics. After having a target, data, and suitable metrics, it is time to analyze it with regard to its type and structure. Then, after reporting the results with more meaningful insights and visualizations, the final step is transforming the business and makes decisions by using these results to improve and optimize existing business processes.

The life cycle of a typical big data project is very similar to that of traditional data projects. Figure 6 depicts the standard big data analytics life cycle, which begins with problem identification and concludes with visualization. The first step is to figure out what data is required for the desired output and which approaches can be used. The

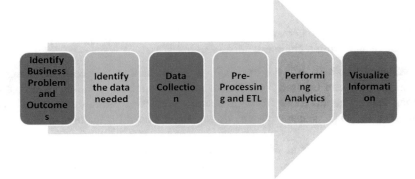

Fig. 6 The big data analytics life cycle [8]

next stage is to determine the data's quality, quantity, format, and sources [8]. The next step is data gathering, which is followed by preprocessing because the data may be in multiple forms or have quality issues. The next crucial step is to use analytics to explain the data and uncover relationships. This step usually includes figuring out what happened and why it happened. Lastly, the final step is visualization of information to get insights and make decisions based on these analyses.

3.1 An Example of Big Data Analytics Architecture for Industry 4.0

The use of relevant Big Data technologies connected to fulfill data gathering, storage, processing, and analysis demands is required for the transformation to Industry 4.0 [77]. The suggested architecture is divided into seven layers, each of which has components linked to certain technical instruments. While each layer is displayed by a rectangle in Fig. 7, dashed rectangles are utilized where applicable to denote components and associated technologies. This method is also used to indicate data flows between layers [76]. These layers are as follows:

1. The Entities/Applications layer: All big data providers and consumers are represented in this layer.
2. The Data Sources layer: This layer includes components like databases, files, email, and Internet services. These components can produce data with low speed and concurrency or high speed and concurrency.
3. The Data Preparation layer: This is the process of extracting data from data sources and transferring it to the data storage layer. Talend is used to combine data from numerous data sources, which is one of the many technologies that may be utilized to accomplish the Data Preparation process.

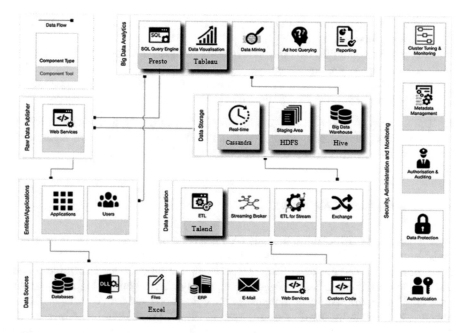

Fig. 7 Architecture of big data for industry 4.0 [77]

4. The Data Storage layer: The Data Storage layer consists of many components that will be utilized in distinct scenarios:

   • For real-time, data streams will be saved in a NoSQL database in real time.
   • The components of the staging area and the Big Data Warehouse (BDW) will preserve data in a more historical viewpoint. Data is saved in the Hadoop Distributed File System (HDFS) and made accessible for usage in the Staging Area component for a set amount of time.

5. The Raw Data Publisher layer: This layer allows web services to be used to download data from the Data Storage layer.
6. The Big Data Analytics layer: Big Data Analytics is made up of elements that make it easier to analyze massive quantities of data and provide access to various data analysis methodologies. The big data analytics components are as follows:

   • Data visualization is a technique for exploring and analyzing data using simple and understandable graphics.
   • Data Mining (also known as Knowledge Discovery) is the process of discovering unusual patterns and insights in data.
   • Ad-hoc Querying allows users to define queries on data interactively while also taking into consideration their analytical needs. Queries are created in real time, frequently depending on the outcomes of earlier data analyses. This

element should provide an interrogation environment that is simple to use and straightforward.
- Reporting compiles data into information briefs to track how various aspects of a company are operating.
- SQL Query Engine provides an interface between the other components in this layer and the Data Storage layer [77].

7. The Security, Administration, and Monitoring layer: Lastly, this layer comprises components that offer fundamental functionality required by other levels and ensure the seamless operation of the complete infrastructure [77].

3.2 Methods and Relations

We can think that artificial intelligence (AI) is a campus having a machine learning (ML) house in it and deep learning (DL) is a room in this house, and the ways from room to out-campus can be interpreted as data mining. So we know that if one has bigger land, it can create a bigger, better campus analogous to big data and AI relation. Hence, the statement that small data cannot be used for AI, ML, or DL is not true, but beaten by big data as depicted in Fig. 8. So, AI can be defined as all attempts to make processes automatic and cognitive. If DL were a fireplace, then data would become its wood [19].

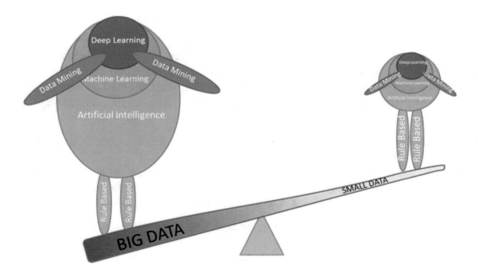

Fig. 8 Big data and small data comparison by showing their relation to artificial intelligence, machine learning, deep learning, data mining, and rule-based methods where big data is more composed of from artificial intelligence to data mining, and on the other hand, small data uses more rule-based methods

3.2.1 Big Data Analytical Methods

BDA enables organizations to make better use of their data and uncover new possibilities [70]. While big data has limitless potential, it is constrained by the availability of big data analytics technology, tools, and capabilities. Big data analytics may be thought of as a subset of the broader process of deriving insights from large amounts of data. Using big data, it is feasible to enhance decision-making and optimize firm productivity. Numerous analytical techniques are used to generate meaning from data [81]. These include the following:

- Descriptive analytics: Through the use of standard reports, ad hoc reports, and alerts, descriptive analytics evaluates data and information in order to identify the current condition of a business situation in such a manner that growth, patterns, and outliers become visible [51].
- Inquisitive analytics: Data analysis to verify or refute business ideas [15].
- Predictive analytics: Prediction and statistical modeling are used to ascertain future possibilities [91].
- Prescriptive analytics: Focuses on optimization and randomized testing in order to understand how businesses might increase service levels while decreasing expenses [51].
- Preemptive analytics: This term refers to the capability of an organization to take preventive steps in reaction to happenings that may have a negative influence on its performance [54] (Fig. 9).

3.3 Big Data Tools

The rising importance of big data and analytics has attracted disruptive technologies' attention. Converting big data to real-world information may aid in decision-making and performance improvement in e-commerce, e-government market intelligence, science and technology, security, smart health, and public safety [16]. Apart from the above paragraph, big data has been significant in a variety of decision-making and forecasting fields, including recommendation systems, business analysis, health care, online display advertising, physicians, transportation, fraud detection, and tourist marketing [68]. Regrettably, sixty percent of big data projects never progressed beyond the prototype level in 2017. As a result of these obstacles, many firms were unclear if investing in big data would enhance their performance [40]. Thus, the platform on which we use big data is critical, as is the data's legitimacy and relevance, its validity, value, volume, variety, and speed.

There are several technologies available for large data analytics, including Spark, Storm, Hadoop, and NoSQL [WK1]. Numerous big data technologies, such as Hadoop, Spark, and others, have accelerated the distribution, interchange, and analysis of enormous volumes of data in the scientific and industrial worlds. Apache Hadoop and, more recently, Spark are well-known frameworks for processing

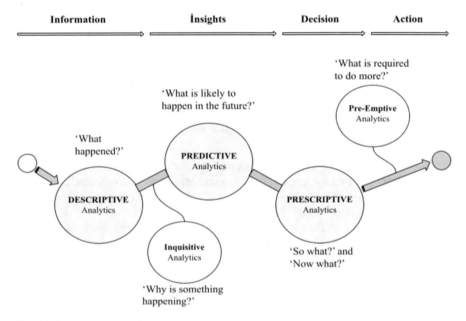

Fig. 9 Big data analytical methods

massive volumes of data based on the MapReduce paradigm. They enable effective application of data mining and machine learning techniques across a range of areas [68].

This section describes various tools and technologies and discusses their application areas.

Apache Hadoop is a free and open-source framework for leveraging simple programming ideas to distribute the processing of large datasets among computer clusters. Apache Hadoop is a software framework that provides an easy-to-use platform for distributed processing of enormous data volumes across computer clusters. Hadoop is designed to scale from a single server to thousands of devices, each of which provides computation and storage capabilities on a local level [9].

Apache Spark is a more recent contender for Hadoop's throne. Additionally, it includes a component called MLlib, which is a machine learning library focusing on clustering, classification, regression, and even data preparation. Due to the capabilities of Spark, batch and streaming analysis may be carried out on the same platform. Spark was intended to solve Hadoop's disadvantage of being unoptimized for iterative algorithms and interactive data analysis, and both of which require executing several operations on the same set of data. Spark is classified as the next generation of distributed computing frameworks due to its memory-intensive architecture, which enables it to process massive amounts of data in memory with a rapid response time [68].

Cassandra encrypts data between clients and nodes in a transparent manner. Client-to-node encryption utilizes the secure socket layer to safeguard data in transit from client machines to a database. Transparent data encryption encrypts data in transit, preventing unauthorized users from accessing the data. The column database Hyper-Table does not provide data encryption during storage. Additionally, data loss occurs in HyperTable as a result of frequent master node failures. Cassandra does not provide join or aggregate operations, although HBase does. Both Cassandra and HBase are capable of processing all types of transactions, albeit HBase has less security features than Cassandra [71].

Apache Storm is an open-source distributed real-time computing solution for big data processing. It is easy to use; it is compatible with any programming environment; and it is helpful for machine learning, ETL, continuous monitoring of activities, distributed remote procedure calls (RPC), and real-time analytics. Due to Storm's scalability, fault tolerance, reliability, speed, and ease of use, it is well suited for real-time data processing [10].

Syncfusion is a commercial solution for managing large amounts of data. It is a huge data platform for Windows. There are a variety of challenges associated with the processing and administration of enormous amounts of data. This platform enables various critical features, including structured and unstructured data query processing and cost-effective storage. Due to the linear scalability of commodity technology, we can store any type of data. Thanks to Syncfusion, these cutting-edge technologies are now available on Windows. We have complete access to the Hadoop framework via this platform. Numerous businesses, including Amazon, Yahoo, Adobe, Facebook, Microsoft, and Hulu, utilize this foundation [10].

Cloudera Data Platform (CDP) is the industry's first corporate data cloud, combining the best of Hortonworks and Cloudera technology. CDP offers robust self-service analytics across hybrid and multi-cloud environments, as well as the sophisticated and extensive security and governance rules required by IT and data professionals. It enables numerous analytic functions to collaborate on the same data at the source, hence removing inefficient and costly data silos. It adheres to stringent enterprise-wide data security, governance, and control standards across all settings. It is entirely open source, with open computing and open storage, providing vendor independence and maximum compatibility [94].

3.4 Applications to Industry 4.0

Industry 4.0 is being implemented with the assistance of modern technologies. Industry 4.0 will be implemented successfully through the use of big data, artificial intelligence (AI), robotics, the Internet of Things (IoT), cloud computing, and 3D printing. The main purpose of these technologies is to collect the data necessary to resolve an issue that arises during manufacturing or other critical services. Because it is the foundation of these technologies, big data is critical to driving development in this fourth industrial revolution. To summarize, big data applications are advantageous for in-process management and productivity enhancement in the automation sector. Complex systems of drivers and intelligent sensors may be easily optimized using the data collected by this technology. Big data is critical for achieving a competitive edge by discovering fundamental manufacturing process issues such as process irregularities, quality discrimination, and energy efficiency waste [49].

The word "big data" is a loaded term with several meanings and connotations. Its emphasis has evolved over time from the features of datasets in respect to current technologies to systems that enable the economically efficient extraction of value from very large volumes of a diverse range of data through high-velocity gathering, discovery, and analysis. Enterprise data is another form of big data that is gaining importance as Industry 4.0 progresses. Enterprises create and manage massive volumes of data: In addition to internal accounting, employee data, and internal communications, regulatory data custody requirements apply. This also applies to institutions, where this sort of data (e.g., administrative, billing, and scheduling information) broadens the pool of possible sources and enables the conduct of research that is not exclusively biological or medical in nature. This is only expected to rise in Industry 4.0, as the emphasis will be on complete data exploitation, bolstered by more data sources and process information. This adds to the external data generated by sold things, consumers, and suppliers/partners, demanding the further deployment and growth of Big Data technologies [3].

Connecting everyday things to the Internet, informally referred to as the Internet of Things (IoT), is widely cited as a critical enabler for developing new intelligent applications and services. The Internet of Things may be founded on the broad deployment of a variety of devices, such as RFID tags, sensors, actuators, and mobile phones, that are capable of communicating with one another and collaborating with other services to accomplish common goals. Numerous Internet of Things technologies have been proposed for deploying context-aware apps in a straightforward manner. However, modern IoT applications, such as ThingSpeak, may be used to monitor and analyze data (e.g., traffic monitoring) with no assurances on transmission and processing. As a result, it may take tens of seconds to feed data from an embedded device into a data analytics cloud and do analysis on the data [60].

Parallel to improvements in IoT platforms, different Big Data technologies have been developed. These technologies may be classified into two categories based on their data processing principles: batch processing technologies and real-time processing technologies. Batch processing technologies, such as Hadoop, are better suited for processing large amounts of data quickly. The concept is to store data first and then process it. Real-time processing tools, such as Storm and S4, have been developed to rapidly analyze data in motion and derive actionable insights. The objective is to enable the development of applications that require real-time or near-real-time processing. These applications should be utilized in environments with rigorous requirements, such as high data rates (over ten thousand events per second) and low latency (less than a few seconds) [60].

With recent improvements in more affordable sensors, improved data collection systems, and faster communication networks in Industry 4.0's Cyber Physical System

(CPS), there is an increasing use of networked physical facilities that will generate a massive amount of data to analyze, dubbed big data. Manufacturing generates the most data and stores it the most efficiently of any industry sector. As a result, big data analytics has grown in importance for CPS and Industry 4.0 [95].

Additionally, big data analytics and application development for Digital Twins, SOA, and virtualization for Industry 4.0.

3.4.1   Cases of Literature on BDA Applications

- A big data analysis study in the field of marketing

The ability to assess vast amounts of consumer datasets has become a critical concern in the modern day. For example, opinion data is increasingly being generated online and made accessible through a range of media, including consumer evaluations, twitter feeds, and blogs. This opinion data demonstrates the consumer's fundamental requirements [50].

Jin et al. [50] sought to contribute to the product design process by doing sentiment analysis on product evaluations on Amazon.com.

Thus, market requirements might be established. Numerous mobile phone brand evaluations are categorized as good, negative, or neutral using the Bayesian analysis approach. After all, the most often mentioned attributes of users' phones were the screen, battery, apps, network, and memory. The three most often mentioned characteristics between 2003 and 2014 were identified through a review of customer comments, and it was discovered that customers mostly commented on the batteries and displays of their phones. Naturally, these meanings have varied from year to year. For instance, in the past, people remarked on photographs shot using their phones. With the advancement of information and communication technology, phones have evolved into devices with Internet connectivity, and surprising apps have been invented and discussed. The research examined the implementation of quality functions to satisfy consumer expectations across four distinct phone brands. Thus, they were able to determine which features of which phone brand were more popular with buyers.

- A big data analysis study in the field of health care

Due of the healthcare implications of IoT, its storage, processing, scalability, and networking capabilities have been curtailed. The combination of cloud computing and IoT enables a new range of storage, processing, scalability, and networking capabilities previously accessible in IoT. On the other hand, big data processing in Industry 4.0 applications continues to be a serious challenge [34].

Elhoseny et al. [34] propose a unique paradigm for Cloud-IoT-based healthcare applications in an integrated Industry 4.0 environment in order to address this issue.

They demonstrated a novel methodology for improving virtual machine (VM) selection in healthcare applications.

To manage large amounts of data, the suggested technique makes use of a hybrid Internet of Things-cloud computing paradigm. Industry 4.0 applications require automated data processing and analysis. The proposed solution aims to enhance healthcare system performance by reducing the time required to execute stakeholders' requests, improving the storage of patients' large data, and providing real-time data collection tools for these applications. The four critical components of the hybrid cloud-IoT architecture are stakeholder devices, stakeholder requests (tasks), cloud broker, and network management. The Genetic Algorithm (GA), the Particle Swarm Optimizer (PSO), and Parallel Particle Swarm Optimization (PPSO) were used to optimize VM selection. For the purpose of estimating the execution time of stakeholder requests, the proposed eligibility function is a mix of three critical criteria: CPU consumption, turnaround time, and wait time. Experiments were done to compare the execution times, data processing speeds, and system efficiency of these three optimizers. Against determine its efficacy, the suggested model was compared to cutting-edge technologies. As a consequence, it was revealed that the suggested model significantly outperformed existing models in terms of overall execution time. Additionally, system efficiency has enhanced by 5.2% in terms of real-time data retrieval [34].

- A big data analysis study in the field of oil and gas industry

Nguyen et al. [69] did a comprehensive evaluation of big data analytics for the fourth industrial revolution in the oil and gas (O&G) industry. As a consequence of the investigation, several issues regarding the industry's productivity rise were identified.

Analytics on big data (BD) is a significant component of the oil and gas (O&G) industry's digitalization. The objective is to improve operational efficiency, facilitate decision-making, and mitigate risk. This is accomplished by the ability to handle and process enormous amounts of data. Additionally, improved seismic data processing enables the industry to have a deeper grasp of BD applications. However, the sector is still slow to accept new technology. This is because critical issues such as defense and cyber security must be considered when integrating with current systems. In some instances when wearable devices are used, physiological and location data create workplace privacy concerns. These shortcomings cast doubt on the practical benefits and feasibility of adding BD into O&D activities. The essay provides an in-depth examination of BD analytics from the perspective of the oil and gas business. Numerous challenges have been highlighted as impeding the adoption of Industry 4.0 technology in the oil and gas sector and inhibiting its successful use.

These issues include the following: When governments employ BD analytics, they are cautious of data privacy considerations. Despite numerous efforts to develop and modernize legal frameworks to ensure greater privacy and security for Industry 4.0, legal gaps between nations continue to exist, posing significant challenges in managing cross-border data. Cybersecurity-related IT concerns must be addressed as part of the BD rollout. Specialized software tools and platforms are necessary to detect and combat cyber-attacks on systems. To decrease the time required for the system to react and reset following a cyber assault, numerous adjustments in management and collaboration between IT and other departments are necessary.

This is an excellent opportunity to address these concerns. Digitization is exemplified through digital twins, wearable computing, artificial intelligence, blockchain, and robotics. After all, by applying these technologies, it will be possible to operate in an integrated manner.

3.5 Computational Aspects of Big Data

Real-time or near real-time usage is an important concept for big data analytics. The value of a decision may be determined by the amount of time it took to make it, for example, accepting or rejecting a decision, such as fraud detection in financial transaction streams, or recommendation systems. The term real time refers to data processing at the speed of a business [8]. For a financial institution that is detecting fraud, real-time means milliseconds, or it means seconds to a retail company doing click-stream analytics.

Batch processing and real-time processing are two paradigms for data processing. In general, high latency is a fundamental characteristic of batch processing applications, whereas low latency is a characteristic of real-time applications. There are also some subdivisions of batch processing such as mini-batch processing which provides moderate latency. To maintain low latency, real-time systems handle incoming data when it arrives without necessarily writing it to a database. Stream processing is a popular type of real-time applications with high volume [8]. On the other hand, complexity and costs are two consequences of implementing real-time solutions.

3.6 Opportunities and Challenges of Big Data Analytics

While big data presents new opportunities to modern civilization, it also presents challenges to data analysts [35]. This section will discuss the opportunities and challenges that come with big data analytics.

3.6.1 Opportunities of BDA

Big data has enormous promise for businesses in terms of launching new ventures, producing new goods and services, and enhancing operations. Big data analytics may provide advantages such as cost savings, improved decision-making, and improved quality of products and services [25].

1. Marketing personalization: Firms can send individualized product/service suggestions, discounts, and other promotional proposals by using big data from many sources [56].
2. Improvement in pricing: Companies can price effectively by utilizing the stream of data existing from customer communications [11].
3. Lowering costs: It is very important for companies to understand their customer profiles. A manual technique was used even before modern technology was applied to research customer behavior and develop new marketing tactics. With the globalization of business and the expansion of knowledge, it has become almost impossible to continue in this way. Multi-channel marketing initiatives are very important today. Big data technologies are used by successful marketers to analyze customer attitude and make strategic decisions [86].
4. Enhanced service to customers: Big data analytics may help customer care representatives comprehend the background of customer problems comprehensively and swiftly resolve them by combining data from numerous communication channels. Big data analytics may also be used to evaluate financial activity in real time, detect potential fraud, and alert customers to possible problems quickly [58].

Here are some examples of how some of the world's most well-known major organizations are utilizing big data:

By analyzing big data generated by customer data, Walmart discovered that when Hurricane Sandy struck in 2004, there was an interesting spike in strawberry Pop Tart sales in addition to natural disaster relief supplies before the hurricane. Thereupon, these products were sent to the stores on the way to the 2012 Frances hurricane and achieved a good sales rate. This is an excellent example of big data analytics [64].

Rolls-Royce uses big data in design, production, and after-sales support services [63].

Netflix can reveal the watching habits of the viewers with the data collected from its millions of viewers. Thus, the company adds millions of new customers to its portfolio every year, with the data it obtains and the content it develops in line with the tastes of the customers [63].

GE assists the oil and gas sector in improving equipment dependability and availability, which leads to increased operational efficiency and production. Real-time monitoring systems send vast volumes of data to central facilities, where it is analyzed using data analytics to determine the state of the equipment. Southwest Airlines uses GE's patented flight efficiency algorithms to evaluate flight and operational data, identifying and prioritizing fuel-saving options [44].

3.6.2 Challenges of BDA

Opportunities are usually accompanied with obstacles. On the one hand, big data offers enormous opportunities, but it also brings significant challenges. Among the issues are data collection, storage, search, sharing, analysis, and visualization [5].

Major issues in big data analysis include data inconsistency and incompleteness, scalability, velocity, and data security [4].

To begin, data analysis should begin with a well-structured set of data. Given the diversity of datasets in big data difficulties, unstructured or semi-structured data processing continues to be a significant barrier. Additionally, the dataset is typically fairly large and is derived from a number of sources. Existing real-world databases are particularly prone to inconsistency, incompleteness, and noise in the data. A range of data preparation techniques can be employed to minimize noise and correct discrepancies [43].

1. Inconsistency and incompleteness: Big data contains information of varying reliability supplied by a growing range of sources. Uncertainty, inaccuracy, and missing values occur much too frequently and must be handled. On the plus side, the breadth and abundance of big data frequently enable the filling in of gaps, resolving arguments, confirming trusted linkages, and discovering hidden correlations and patterns. Even when the problem remedy is applied, certain omissions and errors in the data are almost certain. When conducting data analysis, it is necessary to remedy this weakness and these errors [48].

2. Scalability: Of course, the first worry with big data is its size. For decades, handling massive and ever-growing volumes of data has been a challenge [48].

3. Velocity: The difficulty with velocity is that non-homogeneous data must contend with a high rate of data entry. This results in either the generation of new data or the updating of existing data [17].

4. Security: A lack of security contributes to user reluctance to accept big data. Additionally, it can result in financial losses and damage to a business's reputation. If adequate security protections are not in place, confidential information may be mistakenly conveyed to unwanted recipients. This security risk may be mitigated by creating a strong security management protocol and integrating security solutions such as attack prevention and detection systems, encryption, and firewalls into big data platforms. Blockchain technology, which underpins the Bitcoin cryptocurrency, is a potential tool for securing massive amounts of data. By encrypting data rather than storing it in its original form, blockchain ensures that each data item is unique, timestamped, and tamper-proof, and its applications extend beyond the financial industry due to its enhanced level of data security [58].

References

1. 1 Second—Internet Live Stats (n.d.) Retrieved January 2, 2022, from https://www.internetlive stats.com/one-second/
2. Abawajy J (2015) Comprehensive analysis of big data variety landscape. Int J Parallel Emergent Distrib Syst 30(1):5–14
3. Aceto G, Persico V, Pescapé A (2020) Industry 4.0 and health: internet of things, big data, and cloud computing for healthcare 4.0. J Ind Inf Integr 18:100129

4. Agrawal D, Bernstein P, Bertino E, Davidson S, Dayal U, Franklin M, Gehrke J, Haas L, Halevy A, Han J, Jagadish HV, Labrinidis A, Madden S, Papakonstantinou Y, Patel J, Ramakrishnan R, Ross K, Shahabi C, Widom J et al (2011) Challenges and opportunities with big data 2011-1

5. Ahrens J, Hendrickson B, Long G, Miller S, Ross R, Williams D (2011) Data-intensive science in the US DOE: case studies and future challenges. Comput Sci Eng 13(6):14–24

6. Alper P, Belhajjame K, Goble C, Karagoz P (2013) Small is beautiful: summarizing scientific workflows using semantic annotations. In: 2013 IEEE international congress on big data. IEEE, pp 318–325

7. Anderson A (2015) Statistics for big data for dummies. John Wiley & Sons

8. Ankam V (2016) Big data analytics. Packt Publishing Ltd

9. Anuradha J (2015) A brief introduction on big data 5Vs characteristics and Hadoop technology. Proc Comput Sci 48:319–324

10. Arfat Y, Usman S, Mehmood R, Katib I (2020) Big data tools, technologies, and applications: a survey. In: Smart infrastructure and applications. Springer, Cham, pp 453–490

11. Baker W, Kiewell D, Winkler G (2014) Using big data to make better pricing decisions. In: McKinsey analysis

12. Barnes TJ (2013) Big data, little history. Dialogues Human Geogr 3(3):297–302

13. Beyer MA, Laney D (2012) The importance of "big data": a definition. Stamford, CT

14. Big Data—Google Trends (n.d.) Retrieved December 30, 2021, from https://trends.google.com/trends/explore?date=all&q=Big%20Data

15. Bihani P, Patil ST (2014) A comparative study of data analysis techniques. Int J Emerg Trends Technol Comput Sci 3(2):95–101

16. Chang V (2021) An ethical framework for big data and smart cities. Technol Forecast Soc Chang 165:120559

17. Chen J, Chen Y, Du X, Li C, Lu J, Zhao S, Zhou X (2013) Big data challenge: a data management perspective. Front Comp Sci 7(2):157–164

18. Chen H, Chiang RH, Storey VC (2012) Business intelligence and analytics: from big data to big impact. MIS Quarterly 1165–1188

19. Chollet F (2021) Deep learning with Python. Simon and Schuster

20. Clement J (2020) Global mobile data traffic from 2017 to 2022. Statista. Retrieved January 10, 2022, from https://www.statista.com/statistics/271405/global-mobile-data-traffic-forecast/

21. Cox M, Ellsworth D (1997) Application-controlled demand paging for out-of-core visualization. In: Proceedings. Visualization'97 (Cat. No. 97CB36155). IEEE, pp 235–244

22. Cukier K (2010) Data, data everywhere. Economist 394(8671):3–5

23. Dai HN, Wang H, Xu G, Wan J, Imran M (2020) Big data analytics for manufacturing internet of things: opportunities, challenges and enabling technologies. Enterp Inf Syst 14(9–10):1279–1303

24. DataReportal (n.d.) Digital around the world—datareportal—global digital insights. Retrieved January 7, 2022, from https://datareportal.com/global-digital-overview

25. Davenport T (2014) Big data at work: dispelling the myths, uncovering the opportunities. Harvard Business Review Press

26. Davenport TH, Barth P, Bean R (2012) How 'big data' is different

27. Davis K (2012) Ethics of big data: balancing risk and innovation. O'Reilly Media, Inc

28. Desjardins J (2019) How much data is generated each day? | World Economic Forum. The World Economic Forum. Retrieved January 30, 2022, from https://www.weforum.org/agenda/2019/04/how-much-data-is-generated-each-day-cf4bddf29f/

29. Diebold FX (2003) Big data dynamic factor models for macroeconomic measurement and forecasting. In: Dewatripont M, Hansen LP, Turnovsky S (eds) Advances in economics and econometrics: theory and applications, Eighth World Congress of the Econometric Society, pp 115–122

30. Diebold FX, Cheng X, Diebold S, Foster D, Halperin M, Lohr S, Mashey J, Nickolas T, Pai M, Pospiech M, Schorfheide F, Shin M (2012) A personal perspective on the origin(s) and development of "big data": the phenomenon, the term, and the discipline∗.

31. Dimitrov DV (2016) Medical internet of things and big data in healthcare. Healthc Inf Res 22(3):156–163
32. Domo, Data never sleeps 9.0
33. Dumbill E (2012) Big data now: 2012 edition. O" Reilley Media, Sebastopol
34. Elhoseny M, Abdelaziz A, Salama AS, Riad AM, Muhammad K, Sangaiah AK (2018) A hybrid model of internet of things and cloud computing to manage big data in health services applications. Futur Gener Comput Syst 86:1383–1394
35. Fan J, Han F, Liu H (2014) Challenges of big data analysis. Natl Sci Rev 1(2):293–314
36. Flouris I, Giatrakos N, Deligiannakis A, Garofalakis M, Kamp M, Mock M (2017) Issues in complex event processing: status and prospects in the big data era. J Syst Softw 127:217–236
37. Frank AG, Dalenogare LS, Ayala NF (2019) Industry 4.0 technologies: implementation patterns in manufacturing companies. Int J Prod Econ 210:15–26
38. Gandomi A, Haider M (2015) Beyond the hype: big data concepts, methods, and analytics. Int J Inf Manage 35(2):137–144
39. Gartner (2014) Gartner says the internet of things will transform the data center
40. Ghasemaghaei M (2021) Understanding the impact of big data on firm performance: the necessity of conceptually differentiating among big data characteristics. Int J Inf Manage 57:102055
41. Gobble MM (2013) Big data: the next big thing in innovation. Res Technol Manag 56(1):64–67
42. Groves P, Kayyali B, Knott D, Kuiken SV (2016) The 'big data' revolution in healthcare: accelerating value and innovation
43. Han J, Pei J, Kamber M (2011) Data mining: concepts and techniques. Elsevier
44. How Big Data and the Industrial Internet Can Help Southwest Save $100 Million on Fuel I GE News (2015) General electric. Retrieved January 12, 2022, from https://www.ge.com/news/reports/big-data-industrial-internet-can-help-southwest-save-100-million-fuel
45. How Much Data is Created Every Day? [27 Powerful Stats] SeedScientific (2021) Retrieved January 10, 2022, from https://seedscientific.com/how-much-data-is-created-every-day/
46. Hurwitz JS, Nugent A, Halper F, Kaufman M (2013) Big data for dummies. John Wiley & Sons
47. IBM Cloud Education (2021) Structured versus unstructured data: what's the difference? IBM. Retrieved January 13, 2022, from https://www.ibm.com/cloud/blog/structured-vs-unstructured-data
48. Jagadish HV, Gehrke J, Labrinidis A, Papakonstantinou Y, Patel JM, Ramakrishnan R, Shahabi C (2014) Big data and its technical challenges. Commun ACM 57(7):86–94.7
49. Javaid M, Haleem A, Singh RP, Suman R (2021) Significant applications of big data in industry 4.0. J Indus Integr Manage Innov Entrepreneurship 429–447
50. Jin J, Liu Y, Ji P, Liu H (2016) Understanding big consumer opinion data for market-driven product design. Int J Prod Res 54(10):3019–3041
51. Joseph RC, Johnson NA (2013) Big data and transformational government. IT Prof 15(6):43–48
52. Jun SP, Yoo HS, Choi S (2018) Ten years of research change using Google Trends: from the perspective of big data utilizations and applications. Technol Forecast Soc Chang 130:69–87
53. Kaisler S, Armour F, Espinosa JA, Money W (2013) Big data: issues and challenges moving forward. In: 2013 46th Hawaii international conference on system sciences, 2013, pp 995–1004. https://doi.org/10.1109/HICSS.2013.645
54. Kerr I, Earle J (2013) Prediction, preemption, presumption: How big data threatens big picture privacy. Stan L Rev Online 66:65
55. Khan M, Wu X, Xu X, ve Dou W (2017) Big data challenges and opportunities in the hype of Industry 4.0. In: 2017 IEEE international conference on communications (ICC). IEEE, pp 1–6
56. Kwon O, Lee N, Shin B (2014) Data quality management, data usage experience and acquisition intention of big data analytics. Int J Inf Manage 34(3):387–394
57. Laney D (2001) 3-D data management: controlling data volume, velocity and variety. META Group Res Note (February) 1–4
58. Lee I (2017) Big data: dimensions, evolution, impacts, and challenges. Bus Horiz 60(3):293–303

59. Liao J, Zhao Y, Long S (2014) MRPrePost—A parallel algorithm adapted for mining big data. In: 2014 IEEE workshop on electronics, computer and applications, pp 564–568. https://doi.org/10.1109/IWECA.2014.6845683
60. Malek YN, Kharbouch A, El Khoukhi H, Bakhouya M, De Florio V, El Ouadghiri D, Latré S, Blondia C (2017) On the use of IoT and big data technologies for real-time monitoring and data processing. Proc Comput Sci 113:429–434
61. Manyika J, Chui M, Brown B, Bughin J, Dobbs R, Roxburgh C, Hung Byers A (2011) Big data: the next frontier for innovation, competition, and productivity. McKinsey Global Institute.
62. Marr B (2015) Big data: using SMART big data, analytics and metrics to make better decisions and improve performance. Wiley
63. Marr B (2016) Big data in practice: how 45 successful companies used big data analytics to deliver extraordinary results. Wiley
64. Marr B (2016) The most practical big data use cases Of 2016. Forbes. Retrieved January 10, 2022, from https://www.forbes.com/sites/bernardmarr/2016/08/25/the-most-practical-big-data-use-cases-of-2016/?sh=423d47531625
65. Marx V (2013) The big challenges of big data. Nature 498(7453):255–260
66. Mayer-Schönberger V, Cukier K (2013) Big data: a revolution that will transform how we live, work, and think. Houghton Mifflin Harcourt.
67. Ming Z, Luo C, Gao W, Han R, Yang Q, Wang L, Zhan J (2013) Bdgs: a scalable big data generator suite in big data benchmarking. In: Advancing big data benchmarks. Springer, Cham, pp 138–154
68. Mohamed A, Najafabadi MK, Wah YB, Zaman EAK, Maskat R (2020) The state of the art and taxonomy of big data analytics: view from new big data framework. Artif Intell Rev 53(2):989–1037
69. Nguyen T, Gosine RG, Warrian P (2020) A systematic review of big data analytics for oil and gas industry 4.0. IEEE Access 8:61183–61201
70. Osuszek L, Stanek S, Twardowski Z (2016) Leverage big data analytics for dynamic, informed decisions with advanced case management. J Decis Syst 25(sup1):436–449. https://doi.org/10.1080/12460125.2016.1187401
71. Rao TR, Mitra P, Bhatt R, Goswami A (2019) The big data system, components, tools, and technologies: a survey. Knowl Inf Syst 60(3):1165–1245
72. Reinsel D, Gantz J, Rydning J (n.d.) The digitization of the world from edge to core. Seagate. Retrieved January 30, 2022, from https://www.seagate.com/files/www-content/our-story/trends/files/idc-seagate-dataage-whitepaper.pdf
73. Ross A (2016) The industries of the future. Simon & Schuster, New York, p 320
74. Russom P (2011) Big data analytics. TDWI Best Pract Rep Fourth Q 19(4):1–34
75. SQL Server Team (2013) What does big data mean to you? Microsoft SQL Server Blog. Retrieved January 10, 2022, from https://cloudblogs.microsoft.com/sqlserver/2013/02/08/what-does-big-data-mean-to-you/
76. Santos MY, Oliveira e Sá J, Costa C, Galvão J, Andrade C, Martinho B, Lima FV, Costa E. (2017a) A big data analytics architecture for industry 4.0. In: World conference on information systems and technologies. Springer, Cham, pp 175–184
77. Santos MY, e Sá JO, Andrade C, Lima FV, Costa E, Costa C, Martinho B, Galvão J (2017b) A big data system supporting Bosch Braga industry 4.0 strategy. Int J Inf Manage 37(6):750–760
78. Sebei H, Hadj Taieb MA, Ben Aouicha M (2018) Review of social media analytics process and Big Data pipeline. Soc Netw Anal Min 8:30. https://doi.org/10.1007/s13278-018-0507-0
79. Second (n.d.) Internet live stats. Retrieved January 11, 2022, from https://www.internetlivestats.com/one-second/#traffic-band
80. Sena V, Bhaumik S, Sengupta A, Demirbag M (2019) Big data and performance: what can management research tell us? Br J Manag 30(2):219–228
81. Sivarajah U, Kamal MM, Irani Z, Weerakkody V (2017) Critical analysis of big data challenges and analytical methods. J Bus Res 70:263–286
82. Statista Research Department (2016) Number of IoT devices 2015–2025. Statista. Retrieved January 30, 2022, from https://www.statista.com/statistics/471264/iot-number-of-connected-devices-worldwide/

83. Stegner B (1788 Articles Published) (2020) Memory sizes explained: gigabytes, terabytes, and petabytes in context. MUO. Retrieved January 10, 2022, from https://www.makeuseof.com/tag/memory-sizes-gigabytes-terabytes-petabytes/
84. Streaming Data versus Static Data. SAS help center (n.d.) Retrieved January 7, 2022, from https://documentation.sas.com/doc/en/esptex/5.2/streaming-static.htm#:~:text=The%20oper ational%20difference%20between%20static,database%20before%20it%20is%20processed
85. Sunagar P, Hanumantharaju R, Siddesh GM, Kanavalli A, Srinivasa KG (2020) Influence of big data in smart tourism. In: Hybrid computational intelligence. Academic Press, pp 25–47
86. Suthar S (2018) How to use big data to reduce enterprise costs. SAP Blogs. Retrieved January 12, 2022, from https://blogs.sap.com/2018/02/23/how-to-use-big-data-to-reduce-enterprise-costs/
87. Tu DQ, Kayes ASM, Rahayu W, Nguyen K (2020) IoT streaming data integration from multiple sources. Computing 102(10):2299–2329
88. Venkatraman S, Venkatraman R (2019) Big data security challenges and strategies. AIMS Math 4(3):860–879
89. Verma JP, Agrawal S, Patel B, Patel A (2016) Big data analytics: challenges and applications for text, audio, video, and social media data. Int J Soft Comput Artif Intell Appl (IJSCAI) 5(1):41–51
90. Villars RL, Olofson CW, Eastwood M (2011) Big data: what it is and why you should care. White Paper IDC 14:1–14
91. Waller MA, Fawcett SE (2013) Data science, predictive analytics, and big data: a revolution that will transform supply chain design and management. J Bus Logist 34(2):77–84
92. Wang Y, Kung L, Byrd TA (2018) Big data analytics: understanding its capabilities and potential benefits for healthcare organizations. Technol Forecast Soc Chang 126:3–13
93. Weiss SM, Indurkhya N (1998) Predictive data mining: a practical guide. Morgan
94. www.cloudera.com, 10/12/2021
95. Xu LD, Duan L (2019) Big data for cyber physical systems in industry 4.0: a survey. Enterp Inf Syst 13(2):148–169
96. Zhan J, Han R, Zicari RV (eds) (2016) Big data benchmarks, performance optimization, and emerging hardware: 6th workshop, BPOE 2015, Kohala, HI, USA, August 31-September 4, 2015. Revised selected papers, vol 9495. Springer

Blockchain Technology in Supply Chain Management: Challenge and Future Perspectives

Mahdi Arabian, Mazyar Ghadiri Nejad, and Reza Vatankhah Barenji

Abstract Modern supply chains include multi-layered and geographically separated entities, which leads to the globalization of supply chains and the need for separate regulatory policies. Because of that, the process of managing and controlling information and material flows throughout the supply chain structure has become laborious. Nevertheless, being in the era of information and communication technology, various technologies have been proposed for the facilitation of more efficient supply chain management. One of the cases that have been widely discussed in recent years is the use of blockchain technology as a distributed digital ledger that guarantees the integrity, traceability, transparency, and security of information. In this study, we discuss the overall potential for Blockchain in supply chain management to shed light on this burgeoning subject.

1 Introduction

Recent technologies, most notably those linked with Industry 4.0, are causing substantial disruptions and compelling supply chain management practitioners to build new business strategy models. Blockchains are one of the most promising of these technologies that evolved as a tool to facilitate bitcoin transactions. The blockchain technology idea is an emerging technology that permits the decentralized and unchangeable storing of verifiable data. It has progressively drawn the interest of

M. Arabian
Department of Industrial Engineering, Buali Sina University, Hamedan, Iran

M. Ghadiri Nejad (✉)
Industrial Engineering Department, Girne American University, Kyrenia 99428 TRNC, Turkey
e-mail: mazyarghadirinejad@gau.edu.tr

R. V. Barenji
Department of Engineering, School of Science and Technology, Nottingham Trent University, Nottingham NG11 8NS, UK

© The Author(s), under exclusive license to Springer Nature Singapore Pte Ltd. 2023
A. Azizi and R. V. Barenji (eds.), *Industry 4.0*, Emerging Trends in Mechatronics,
https://doi.org/10.1007/978-981-19-2012-7_9

several businesses during the last few years. The supply chain management community is gradually becoming aware of the enormous impact that blockchain technology might have on their sector.

2 Supply Chain Definition and Structure

Following the rise of global resources in the early 1980s, the concepts of supply chain and supply chain management were introduced to describe the complexities of business-to-customer and/or business-to-business networks. This supply chain, which was initially developed as a concept of purchase and logistics, is now known as the traditional supply chain [1].

The traditional supply chain has evolved over the last four decades. Nowadays, supply chains are inherently more complex and include different stages, high diversity of inputs and outputs, multi-layered and geographically dispersed entities and large numbers of stakeholders that are arduous to manage. All of these complexities made the one, which is known as the modern supply chain [2, 3].

Generally, the supply chain includes all the stages and processes that play a role in creating consumers' satisfaction in fulfilling their orders. These stages often include suppliers, manufacturers, distributors, retailers, and consumer [4]. Figure 1 illustrates the material and information flows in the supply chains.

During these stages, and in order to transfer the products or services to consumers from the suppliers, the flow of materials and information are constantly established between these stages, which can be forward and backward. This multiplies the importance of making good decisions at the operational, tactical and strategic levels in the supply chain. In fact, optimizing the activities performed in the operational stages, including sourcing, storage, production process, transportation process, as well as the flow of information and materials between supply chain stages, are what supply chain management refers to. [5–7]. Therefore, designing the supply chain structure, managing it, and the flows within it through inter-organizational relations are the main aspects of supply chain management [8].

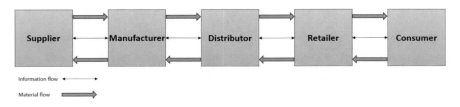

Fig. 1 Flow of material and information in supply chain structure

In general, supply chain management affects the long-term performance of the entire supply chain and achieving supply chain goals is possible through its management [8]. According to this concept, supply chain management goals can be divided into the following main sections:

2.1 Maximizing the Surplus

In general, the first primary goal of supply chain management is to create added value for stakeholders across the supply chain, which means higher profitability and ultimately maximizing surplus. It should be noted that the revenue from a supply chain is the difference between the supply price of services and products to the consumer at the end of the supply chain, and all costs incurred during the supply chain process [2, 4, 9]. Therefore, maximizing supply chain surplus is only possible through optimizing supply chain management.

2.2 Maximizing Consumer Satisfaction

In a competitive global market environment, companies must respond to the rapidly changing needs and demands of consumers [9]. Therefore, more effective response to consumer needs and the achievement to consumers' satisfaction are of the ultimate goals in supply chain management that requires optimal design of supply chain structure [10].

3 Managing Information Flow

The supply chain includes all the steps and activities performed to deliver the goods to the end user as well as the information flows created in between [11]. One of the most important factors in designing an integrated structure for the supply chain is the sharing of information on the processes, exchanges and products that are done in the stages between them. This, creates the flow of information in the supply chain, which is one of the factors that a Supply is determined based on it [7–10]. Therefore, it is important to know that this flow must be established and information must be shared in the supply chain. This information is mainly divided into the following six categories [12].

1. Product or service
2. Process
3. Inventory
4. Source
5. Order
6. Planning.

Fig. 2 Information flows between milk supply chain stages

Acquiring this information in a useful way helps to make effective and successful decisions in the supply chain [13]. More accurate classification varies depending on the product. For example, the flow of information between the stages of the dairy supply chain can be seen in Fig. 2.

To manage the supply chain, it is necessary to create effective information flows in it. Therefore, supply chain management is not possible without managing its information system. Therefore, information system is one of the main factors affecting the performance of the supply chain, which is needed to develop the competitive advantage of the supply chain [8]. That is why timely sharing of information in the supply chain is one of the key elements in achieving a responsive and successful supply chain [9].

In short, one of the main pillars of supply chain management is information sharing and information flow. Therefore, in a closer look, improving decisions, quality of responses, level of services or operations, and creating a competitive advantage, through the flow of information in the supply chain can be effective. On the other hand, the impact of this information on the supply chain depends on the quality of the information, and in order to improve the performance of the supply chain, the information must be of good quality. The quality of information is determined by some indicators like accuracy, being on time, and adequacy of the information. On the other hand, distorted information flow can confuse supply chain members in decision-making, which ultimately has a negative and significant effect on coordination between supply chain steps [14].

But in general and critically, data and information in the supply chain face many problems and they can be considered as supply chain challenges, which are addressed below.

Supply chains currently use and depend on centralized information management systems, especially in the financial, food, health, and education sectors. This means that third party intermediaries in the supply chain, including the processes of transactions, decision-making, and the use of information, have high authority and storage is controlled by them [15]. Relying on centralized information management systems poses threats such as data integrity in the supply chain, during which it creates threats such as fraud, inefficient transactions, theft and mistrust in the supply chain [2, 15].

Data manipulation in the supply chain and data insecurity is also one of the main problems in supply chain information management [2]. Transparency and visibility of data can also be mentioned as a problem in supply chain information management.

One of the key points in reviewing and evaluating supply chain performance is created through the availability of transaction information for participants and even external observers in the supply chain. Therefore, current markets need to resolve ambiguities and transparent information management at every stage of the supply chain [16].

4 Materials Flow in Supply Chain

Supply chain is a regular set of different activities that start from receiving raw materials and end with the delivery of the final product to the consumer [17]. To do this, as mentioned earlier, materials move between stages of the supply chain. Therefore, material flow between supply chain stages is one of the main factors [10].

In a traditional linear supply chain, there is only one stream of linear material, which the final product is disposed at the end of its useful life. These chains generally use the take-make-dispose model. However, in a circular supply chain, the situation is different where the used products can be resold in the market or reused by the consumer [1, 18]. Therefore, two types of material flow are determined in a circular supply chain. The first is a linear loop that is related to the direct flow of materials along the supply chain and the second is a closed loop that is related to reverse flows along the supply chain that occur for various reasons such as recycling or reconstruction [16, 18]. Examples of these two types of material flows are given in Fig. 3, which is related to the closed loop supply chain, and includes both types of material flows.

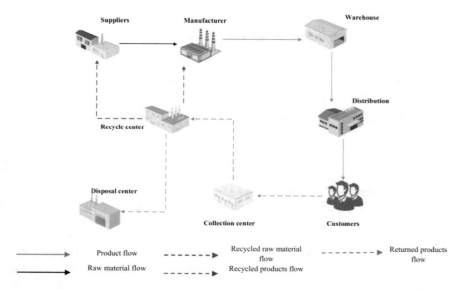

Fig. 3 Material flows between supply chain stages

Materials in the supply chain are not always tangible. Material flow in the supply chain can also refer to intangible material flow. One way to describe supply chain management is an integrated solution for planning and controlling materials and other factors such as information and operations from the first stage of the supply chain (supplier) to the last stage (consumer) [19]. Therefore, the flow of materials in the supply chain has a key role that with its development, along with other factors, better supply chain performance can be achieved [14]. In addition to information and capital flow management, one of the requirements for achieving a sustainable supply chain and what this chain refers to is material flow management [20].

Material flow analysis in the supply chain and the systematic evaluation of material reserves and flows in a system, defined in place and time. As Courtonne points out, this assessment helps to bridge the geographical gap between producers and consumers and can provide a good insight into their shared responsibility [21]. Material flow assessment can also be used to optimize the supply chain network in two dimensions of costs incurred and environmental impacts [22].

One of the challenges we face in the flow of materials along the supply chain is poor product traceability, which leads to theft and smuggling, although the ability to track products can be traced by following the processes performed by all stakeholders [3].

5   Problems

5.1   Information Flow Problems

5.1.1   Data Integration Among the Stages

With the development of information technology (IT) tools and introduction of Industry 4.0 technologies, supply chains are becoming much more complex and geographically dispersed systems [2, 16]. In such systems, information management will not be easy and current information sharing mechanisms in various industries lag behind in terms of new information and communications technologies (ICT) innovations [23].

Transforming the traditional supply chain into a platform that shares centralized data (which may cause data fraud) in the supply chain, and integrate information within the supply chain, have proven to be the methods that in addition to facilitates cooperation in the supply chain, it also affects its performance [24, 25].

Organizations involved in a supply chain consider information and its accuracy to be one of the most important and effective factor in their performance, but they often object to providing and using information when they do not trust each other. Even when they have access to information from their supply chain partners, there is a lack of trust [26]. These relationships between supply chain partners, the management of these relationships and integrated management of the related activities, and internal and external flows of the companies involved in the supply chain to create value for stakeholders. These are what supply chain management refers to, in which one of its main challenges is the integration of IT [2, 8]. Therefore, companies are looking to use new methods, and replace and implement new technologies in addition to the security perspective. In addition, they provide them with capabilities that can ensure the integrity of the data flow, which based on them, they can make better and more reliable plans and forecasts [26].

Current technologies in supply chain, such as electronic data interchange (EDI) or other similar technologies, have created opportunities for companies to move to paperless transactions instead of paper-based transactions. These technologies add new digital capabilities to the supply chain through the ability to transfer documents from computer to computer, but by adopting EDI technology in the supply chain, data processing and exchange of information are controlled by separate systems [16]. Hence, using this technology and centralized data management, supply chain suffers from a lack of data integration in different and separate information systems, which most likely has a significant effect on management costs. In addition, a trusted environment must be created and replaced for the flow of information and data integration among the stages of the supply chain [15, 16].

Finally, according to Lim et al. [25], data integration in supply chains can be ensured by using modern technologies, which is mentioned by ensuring fixed information (on raw materials, processes, and operators), items' flow information (changes in different locations), and changes in ownership information that can be collected according to actual needs.

5.1.2 Data Visibility

One of the key parameters in supply chain evaluation is visibility or transparency, which means easy access to supply chain data and availability of data for exchange parties, external observers, and companies involved in the supply chain. Therefore, especially in competitive, fragmented, and complex markets, many companies are looking to increase supply chain data visibility with new and emerging technologies [16, 27]. According to Pournader et al. [28] transparency tries to shed light on the 'how' aspects for example, how a product is sourced, how it is processed by suppliers, how it is handled while being transported, etc.

As mentioned in the previous section, information management system in most sectors such as financial, food, healthcare, and education supply chain is centralized, and a stakeholder easily controls the entire system. Hence, in a centralized supply chain, the system owner or provider can restrict data sharing system. These barriers, created by human factors, reduce the transparency of data and ultimately the transparency of the entire supply chain [15, 23]. These centralized information management systems, which are responsible for sharing information, are considered as a threat to data availability and consecutively, supply chain suffers from the problem of data transparency [15].

Increasing productivity, reducing costs, and providing better services are the main elements that cause by increasing transparency in the supply chain. Thus, it can be noted that transparency is one of the effective factors in improving supply chain performance [16].

Bai and Sarkis [27] proposed that sustainable transparency in the supply chain includes three dimensions. In the first dimension, it points out that companies and partners involved in the supply chain should participate in the network and share their information, which is a key challenge in the supply chain. However, when organizations and companies involved in the supply chain do share the information that is often stored and kept within organizations, it will be beneficial for both groups of organizations involved in the supply chain and customers. These organizations benefit from increasing efficiency in the supply chain and building mutual trust in the network and developing relationships with their partners. Nevertheless, for the second group, timely sharing of information leads to increased customers' awareness of different aspects of the value of products and services [29]. Examples include knowledge of the origin, quality of used raw materials, the conditions of work done in the supply chain, or the sustainability of production processes [2, 29].

Of course, from an environmental point of view, this subject is also important considering the expansion of supply chains and their increasing rate of globalization. To ensure environmental sustainability and social responsibilities, it is necessary to develop the visible and transparent operations [28].

5.1.3   Data Safety

The possibility of data manipulation in the supply chain, which is a sign of lack of data safety in it, is one of the main concerns of the supply chain. To solve such problem and create immutability in supply chain's data, modern technologies can be used, even though humans still use technologies that can record incorrect data [2]. Therefore, it is said that some operations performed in the traditional supply chain have security problems [3].

Information in the supply chain partners' interrelationships is recognized as a key asset, and for maintaining a competitive advantage, controlled access and proper sharing are essential in the long run [30]. Despite the high benefits of it, information sharing is mostly limited by companies' concerns about their information security and privacy. Hence, there is a lack of trust between companies involved in a supply chain, especially when they have access to each other's information. For this reason, it is possible to intentionally or erroneously send false or misleading information, which will not reflect the real data, and companies are looking to use and implement technologies that enable them to share data securely and ensure the accuracy of data [26, 30].

With the centralized logic commonly used in the supply chain, a central node that has a centralized authority manages each transaction. Such a central node, as a third party, verifies the accuracy and security of the information as well as the task of securely storing and collecting data [15, 16]. In addition, in a centralized ecosystem, the risk of one device being compromised and the possibility of all devices connected to the system being affected, which is a security challenge for supply chains with centralized information management [31]. A more reliable ecosystem than this centralized logic between supply chain stages should be considered and new management methods should be developed to protect supply chain data security [15, 32]. In a decentralized logic, there is no central and unified authority to be referred and used to eliminate intermediaries and eliminate the possibility of data manipulation in the supply chain [16, 25].

The use of emerging technologies to ensure the safety and accuracy of data leads to reduced costs to prevent deliberate and desirable changes which increase supply chain risks [2]. Moreover, increasing trust in transactions between supply chain partners and the end consumer, especially in scenarios where supply chain partners are involved in several layers and do not know each other well, is one of the key elements that the security of supply chain data is effective on it [33]. In addition, the analysis of various aspects of information security indicates that formal, informal, and technical aspects affect the performance of the supply chain in inter-organizational and intra-organizational settings [30].

5.2 Track and Tracing

As mentioned earlier, supply chains are becoming a complex structure with diverse goals, including diversity and a large number of inputs and outputs. The stakeholders are difficult to manage and actually, it faces with many challenges [3].

One of the main challenges in the supply chain is track and tracing of goods across the entire supply chain. Consumers need more traceable system and higher awareness of the origin of products by manufacturers and retailers. However, because of supply chain complexities and the material flow in the extended networks of the supply chain, it is difficult to define the origin, trace defective products to the origin, and detect the safety of products from the beginning to the end of the supply chain, especially in pharmaceutical, agriculture, and food industries [16, 34].

The purpose of supply chain track and tracing is to answer what, where, and when aspects of questions about supply chain inventory transfer [28]. It is said that by creating a proper track and tracing system, it is possible to reduce and minimize the production and distribution of poor quality or unsafe products in the supply chain [15]. Although product traceability in supply chain creates many benefits, but it is an emerging concept that is rarely implemented and there is not enough awareness of the benefits of product track and tracing and consensus about it among supply chain stakeholders. Moreover, rules and regulations about traceability have not been set yet [24]. Therefore, consumers are unable to verify the authenticity of the received product; they are forced to trust the printed certification logo on the product, and cannot verify the integrity of this certificate which requires intense auditing [15].

Traceability enables the observation of supply chain operations and offers insight into market products' custody, authenticity, origin, and integrity. Thus, ensuring efficient traceability is critical for supply chain management [29]. As product reliability and quality requirements increase, the ability to track and trace products has become increasingly vital in the manufacturing business [24]. As Montecchi et al. [29] noted, investments in traceability are frequently motivated by the need to comply with legal and regulatory requirements (e.g., traceability of food items), to improve inventory management, to increase product safety, or to tighten vertical coordination in non-integrated supply chains. Apart from these obvious benefits, many businesses are realizing the extra information value of improved supply chain traceability.

Traceability capability is increased by tracking all processes performed by stakeholders throughout the supply chain, which requires recording the logistics of all products from the very beginning of the supply chain and tracking products throughout the stages and throughout their entire life cycle [3, 16, 24].

6   Blockchain and Distributer Ledger

Blockchain is a new decentralized technology which is a digital ledger distributed throughout a peer-to-peer network and can be used to record and store information and transactions in a secure and unchangeable manner [35]. This technology is highly interoperable and can be a valuable help to the company in achieving Triple Bottom Line (TBL) goals, when it is combined with other technologies developments [36]. Depending on usage and technology applications, blockchain design can form public or private networks and ledgers [2]. As shown in Fig. 4, blockchain technology contains six layers as the following:

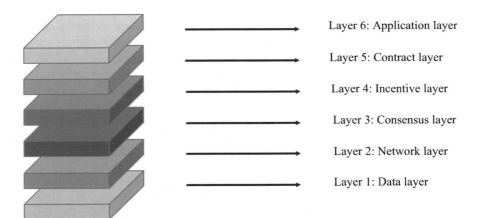

Fig. 4 Structure of blockchain technology

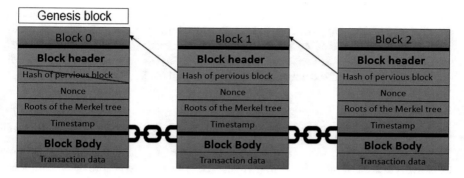

Fig. 5 Structure of blocks in blockchain technology

6.1 Data Layer

The first layer is the data layer, which contains the data stored in the blocks and ensures data integrity. In this layer, the nodes store transactions data received in a specific period in a block. The data will be in a cryptography form using the different hash functions [37]. The hash functions are one-way functions for obtaining input data with using the hashed output. Such functions are not possible to be performed by normal computers; it requires so much computing power, which is not easy to reach [38].

The hash which is generated in the block, based on the transaction records, is also stored in the next block, which means a small change in records of one block influences the whole chain that will be invalid and this makes blockchain completely immutable [37]. Then the exact time that block is generated, will be recorded in timestamp. Finally, blocks will be chained together in a linear and chronological form [39]. The created blocks contain two sections of blocks header and block body (see Fig. 5).

Block header mainly contains the hash of the pervious block and the hash that was generate for this block. The hash is a random number, which is created in an appropriate pattern, called nonce, and can prevent the reply attack and roots of the Merkel tree [40, 41]. The block body contains the data of one or more transaction that have been done [37]. A chain consists of many blocks that are inextricably chained together. The first block in such chain is the genesis block, in which there is no hash of previous block in it and the hash value is equal to 0 [40, 41].

6.2 Network Layer

The network layer includes various protocols for data transfer, distribution, and verification of transactions that are transmitted through the peer-to-peer network in which the nodes are connected at the same level and there is no central node. Therefore, the network is completely decentralized, data moves freely between nodes, and each node can be disconnected or added to the network at any time [42, 43]. Since the next layers, including the consensus and application layers, rely on this layer and its data, the importance of the network layer is very high [44].

6.3   Consensus Layer

Consensus is a set of rules that is implemented by this layer to ensure the generation of valid transactions/blocks and must be followed by each participant [37]. This layer plays a key role in the blockchain technology and the algorithms in this layer has the task of deciding how to verify a block based on the vote of the nodes in the previous layer and make an agreement between them. With the help of this layer, blockchain can be fully decentralized and solve the problem of trust and transparency [35, 45]. In fact, this layer plays an important role in maintaining blockchains' reliability [37]. The algorithms located in this layer can be divided into two main groups:

I.   Proof-based algorithms, which work in such a way that node in the network layer, must prove that they have better conditions for data validation than other nodes. A perfect example for this group is proof of work consensus mechanism.
II. Vote-based algorithms, that need nodes in network so that they can exchange verifying data about a new transaction or block before the final poll. A perfect example for this group is proof of stake consensus mechanism [46].

6.4   Incentive Layer

Blockchain systems frequently employ an incentive structure to maintain the network's majority's integrity. This incentive manifests itself as a block reward for the node that successfully adds a new block to the chain [47]. This layer is mostly utilized for cryptocurrencies, and as Lu [48] said, it attempts to encourage resource sharing in order to boost collective intelligence and collaborative communication. Additionally, this layer incorporates a system for the issuance and distribution of incentives [47].

6.5   Contract Layer

This layer comprises scripts, algorithms, computerized procedures, and smart contracts for the purpose of recording complicated transactions [49]. The term "smart contracts" and its implementation on the blockchain were initially characterized as "systems that automatically transfer digital assets in accordance with predefined rules" [50].

The script code defines the terms and conditions of smart contracts, and when such terms are formed, the contract is immediately executed. The scripts on this layer are used to execute incentive and consensus processes [51].

6.6 Application Layer

In this layer, which is the last layer of the blockchain technology, the results of the actions performed in the previous layers are displayed in the form of an application that can exist in large numbers of industries. These applications that are in various purposes, use blockchain, smart contracts, and cryptocurrencies, and they can be divided into the following groups [52].

6.6.1 Financial Application

The first use of blockchain, which is very popular among the people, Bitcoin, was in the financial sector. It is a solution method to deal with the problem of double spending using a peer-to-peer network. In general, with the help of blockchain technology, Satoshi Nakamoto designed a cryptocurrency that is a solution for security issues in centralized systems. In Bitcoins whitepaper, completely non-reversible transactions are not possible, since financial institutions cannot avoid mediating disputes [53]. The financial sector of blockchain application has gained more attentions so far, and is divided into the following categories.

- Financial services
- Enterprise transformation
- Peer-to-peer financial market
- Risk management.

6.6.2 Non-financial Applications

The public is not fully aware of the capacity of blockchain for non-financial applications that were more known after the blockchain 2.0 and blockchain 3.0. The non-financial applications are divided into the following section [54].

- Internet-of-things
- Security and privacy
- Public and social service
- Reputation system.

7 Types of Blockchain

As mentioned above, blockchain technology can be defined as a public or private ledger that is executed and shared among the agents participate in blockchain [2].

7.1 Public Blockchain

Public blockchains are one of the main parts of blockchains that are implemented through a peer-to-peer network and data is stored publicly as a public ledger [35, 55]. In public blockchains, anyone can add to the network and join the blockchain with anonymity [23]. The public ledger is open to anyone who wants to use open source software, and all participants can conduct transactions, participate in the consensus process, and add new blocks to the chain, but they do not have the ability to modify existing blocks [40, 56].

Almost anybody may use a wallet to access a hash sum that is stored in a blockchain and acts as a unique key (mapped to a specific data or document). As a result, anybody may verify the data's origin and correctness using a cryptographic hash function [26]. The fact that everyone may access and contribute to the public blockchain renders it entirely transparent.

Without the requirement for a trusted third party, public and financial blockchains such as Bitcoin or Ethereum allow all contributors to participate in the control mechanism and agree on a single state of data. The attained consensus is predicated on the majority of participants being honest. Decentralization is frequently used to refer to the practice of delegating control [47].

A public blockchain is totally decentralized, allowing anybody to query messages and record transactions [41]. The absence of trustworthy entities in a distributed system attracts a large number of users in academia and industry [47]. Public blockchains are more suited to cryptocurrency transactions. These blockchains, such as Ethereum and Bitcoin, are limited to processing 3–30 transactions per second and need a significant amount of energy and time to verify blocks [56].

7.2 Private Blockchain

Private blockchains are another type of blockchains in which there is no anonymity and the parties are familiar with and know each other [2]. In private or permissioned blockchains, information can only be provided to a specific group and only an authorized group is allowed to accept the changes of it [46]. In private blockchains there are restrictions for people who intend to participate in the network and they need an invitation that must be approved either by the set of applied rules or by the network starter. Thus, they are able to participate in the network [40]. These blockchains are generally managed by a single entity that the ability of participation of nodes in the network is depended on its permission [37].

Unlike public blockchains which require the entire network to reach a consensus on the state of transactions, in private blockchains, only a number of approved nodes are allowed to participate in the consensus process [56, 57]. The activities of the participants in the network in private blockchains can be limited, so that only certain participants have the ability to carry out certain transactions, which leads to more privacy in the network [40].

Both public and private blockchains are shared between their users. Therefore, without the need for third party authorization, they can record all transactions on a peer-to-peer network [57]. Nevertheless, private and permissioned blockchains have some privileges over public blockchains because of its characteristics and suitable option for industrial and business applications [58]. These characteristics are governance structure, private transactions, and an authentication process, according to Nasir et al. [37].

With fewer network known nodes in private blockchains, transactions are processed faster, consensus is more efficiently, and it would be easier to apply changes in the network [58]. These blockchains have stronger data privacy and information and transactions are more secure because the network is not available to external entities [57, 58]. Private blockchains are also less likely to be attacked (e.g., Sybil attack and 51% attack), as all the nodes that participate in the network have already been validated [58].

8 Discussion and Implications

Today's supply chains work at scale in the majority of circumstances without the use of blockchain technology. Nonetheless, technology has enthused the information technology and supply chain industries. Additionally, it has sparked several publications and spurred big IT companies and start-ups to launch interesting test initiatives. Walmart, for example, tried an application that tracks pigs in China and produce in the United States in order to authenticate transactions and ensure record keeping is accurate and efficient.

The authors are unaware of any large-scale supply chain applications to date, which raises an important question: Is blockchain technology capable of enhancing the value of supply chains?

As most practitioners are aware, many of today's supply chains include high-quality data that they can send at near-real-time speeds between supply chain layers. To determine the value proposition of blockchain technology for the supply chain sector, we need examine three areas where it may offer value:

1. Elimination of inefficient manual processes. While supply chains are capable of handling enormous, complicated data sets, many of their procedures, particularly

those at the lowest supply tiers, are sluggish and largely manual—as is still the case in the shipping sector.

2. Improving traceability. Change is already occurring as a result of growing regulatory and consumer demand for provenance information. Additionally, enhanced traceability offers value by lowering the significant expenses associated with quality issues, such as recalls, reputational harm, or revenue loss from black- or gray-market items. Simplifying a complicated supply chain creates additional chances for value generation.

3. Reducing the cost of supply chain IT transactions. At the moment, this benefit is just theoretical. Bitcoin compensates those who validate blocks or transactions and demands those proposing additional blocks to include a fee in their request. Such a penalty would very certainly be prohibitively expensive in supply chains, owing to their enormous scale. For example, a single automaker will generally issue around 10 billion call-offs to its tier-one suppliers over a 90-day period. Additionally, when all of those transactions are combined, the need for data storage, a critical component of blockchain's distributed-ledger architecture, would greatly increase. Additionally, several copies of data sets would be impracticable in a supply chain setting, particularly in permissionless blockchains.

When a business decides to implement blockchain technology for its supply chain, it must first determine the sort of blockchain it will develop. Recall that the bitcoin strategy is based on a permissionless blockchain filled by unknown and untrusted parties. It is open source and relies on a consensus verification methodology to build confidence in individual blocks. These blockchains lack a central database and central governance.

In contrast, the stakeholders in the majority of supply chains are recognized and trusted. Furthermore, the supply chain industry is unlikely to adopt open access since its users are averse to disclosing private information such as demand, capacity, orders, pricing, and margins to unknown parties at all points along the value chain. This means that the majority of supply chain blockchains would need to be permissioned, with access controlled centrally and limited to known parties having access to certain data segments. This technique theoretically allows for public or private validation of each proposed block. However, we believe it is improbable that proposed blocks in the supply chain world will ever be verified publicly once all stakeholders are known. In the shipping industry, for example, just a few well-known parties—haulers, ports, customs, and shipping lines—are responsible for authenticating each block. When the number of trusted parties is small, the need for independent validation of public-domain consensus protocols is minimal.

9 Conclusion

We argue that while blockchain technology may eventually be beneficial for some types of supply chains, it is not currently ready for widespread deployment. This conclusion is based on the following:

- To yet, blockchain pilots have not established the technology's distinctive value in the supply chain industry.
- Blockchain technology is not yet capable of capturing data across a large number of untrusted parties.
- Providing complete transparency or traceability may be accomplished in a variety of ways, not simply through blockchain technology.
- The cost of establishing and operating a blockchain is unknown at the moment, with few standards in place.

The capacity gap between blockchain's existing capabilities and the capacity required by supply chains is huge.

Blockchain technology can provide trust, transparency, and traceability to supply chains when actors are unknown or untrustworthy. These supply chains are nearly always complicated, multi-tiered, and include a large number of partners, and they operate in a regulated environment that requires a greater level of traceability.

However, for supply chains with well-established and trustworthy participants, a centralized database strategy is frequently more than enough. This is not to say that all of these supply chains are now end-to-end; in fact, many of them rely on siloed databases with minimal traceability. Since a result, many of these supply chains do not require blockchain technology to address these concerns, as they may harness current technologies that are more suited to their high-volume transactions, either alone or in collaboration with partners.

It is premature to quantify the costs of running blockchain technology in the supply chain and to compare them to the expenses of other technologies. Without a doubt, IT firms will be eager to share this information.

The value proposition, on the other hand, must be unambiguous. How efficient are internal transactional processes? How much may end-product failures, recalls, and lawsuits cost? Would a consumer pay a premium for a product that ensures supply chain transparency? These are the kinds of issues that should be explored when evaluating the usage of blockchain in supply chains.

References

1. De Angelis R, Howard M, Miemczyk J (2018) Supply chain management and the circular economy: towards the circular supply chain. Prod Plan Control 29(6):425–437
2. Saberi S, Kouhizadeh M, Sarkis J, Shen L (2019) Blockchain technology and its relationships to sustainable supply chain management. Int J Prod Res 57(7):2117–2135

3. Budak A, Coban V (2021) Evaluation of the impact of blockchain technology on supply chain using cognitive maps. Expert Syst Appl 184:115455
4. Ben-Daya M, Hassini E, Bahroun Z (2019) Internet of things and supply chain management: a literature review. Int J Prod Res 57(15–16):4719–4742
5. Mubarik MS, Kusi-Sarpong S, Govindan K, Khan SA, Oyedijo A (2021) Supply chain mapping: a proposed construct. Int J Prod Res 1–17
6. Lejarza F, Pistikopoulos I, Baldea M (2021) A scalable real-time solution strategy for supply chain management of fresh produce: a Mexico-to-United States cross border study. Int J Prod Econ 240:108212
7. Karimi M, Zaerpour N (2021) Put your money where your forecast is: supply chain collaborative forecasting with cost-function-based prediction markets. Eur J Oper Res
8. Severino MR, Godinho Filho M (2019) POLCA system for supply chain management: simulation in the automotive industry. J Intell Manuf 30(3):1271–1289
9. Dadouchi C, Agard B (2021) Recommender systems as an agility enabler in supply chain management. J Intell Manuf 32(5):1229–1248
10. Wang G, Huang SH, Dismukes JP (2005) Manufacturing supply chain design and evaluation. Int J Adv Manuf Technol 25(1–2):93–100
11. de Souza Henriques R (2019) Multi-agent system approach applied to a manufacturer's supply chain using global objective function and learning concepts. J Intell Manuf 30(3):1009–1019
12. Viet NQ, Behdani B, Bloemhof J (2018) The value of information in supply chain decisions: a review of the literature and research agenda. Comput Ind Eng 120:68–82
13. Aydoğan S, Okudan Kremer GE, Akay D (2021) Linguistic summarization to support supply network decisions. J Intell Manuf 32(6):1573–1586
14. Nimeh HA, Abdallah AB, Sweis R (2018) Lean supply chain management practices and performance: empirical evidence from manufacturing companies. Int J Supply Chain Manage 7(1):1–15
15. Azzi R, Chamoun RK, Sokhn M (2019) The power of a blockchain-based supply chain. Comput Ind Eng 135:582–592
16. Centobelli P, Cerchione R, Del Vecchio P, Oropallo E, Secundo G (2021) Blockchain technology for bridging trust, traceability and transparency in circular supply chain. Inf Manage 103508
17. Gupta S, Drave VA, Bag S, Luo Z (2019) Leveraging smart supply chain and information system agility for supply chain flexibility. Inf Syst Front 21(3):547–564
18. Khompratraporn C (2021) Circular supply chain management. In: An introduction to circular economy. Springer, Singapore, pp 419–435
19. Di Vaio A, Varriale L (2020) Blockchain technology in supply chain management for sustainable performance: evidence from the airport industry. Int J Inf Manage 52:102014
20. Mani V, Gunasekaran A, Delgado C (2018) Enhancing supply chain performance through supplier social sustainability: an emerging economy perspective. Int J Prod Econ 195:259–272
21. Courtonne JY, Alapetite J, Longaretti PY, Dupr´e D, Prados E (2015) Downscaling material flow analysis: the case of the cereal supply chain in France. Ecol Econ J Int Soc Ecol Econ 118:67–80
22. Herczeg G, Akkerman R, Hauschild MZ (2018) Supply chain collaboration in industrial symbiosis networks. J Clean Prod 171:1058–1067
23. Li J, Maiti A, Springer M, Gray T (2020) Blockchain for supply chain quality management: challenges and opportunities in context of open manufacturing and industrial internet of things. Int J Comput Integr Manuf 33(12):1321–1355
24. Tsai YS, Hung WH (2021) A low-cost intelligent tracking system for clothing manufacturers. J Intell Manuf 1–19
25. Lim MK, Li Y, Wang C, Tseng ML (2021) A literature review of blockchain technology applications in supply chains: a comprehensive analysis of themes, methodologies and industries. Comput Ind Eng 154:107133
26. Longo F, Nicoletti L, Padovano A, d'Atri G, Forte M (2019) Blockchain-enabled supply chain: an experimental study. Comput Ind Eng 136:57–69

27. Bai C, Sarkis J (2020) A supply chain transparency and sustainability technology appraisal model for blockchain technology. Int J Prod Res 58(7):2142–2162
28. Pournader M, Shi Y, Seuring S, Koh SL (2020) Blockchain applications in supply chains, transport and logistics: a systematic review of the literature. Int J Prod Res 58(7):2063–2081
29. Montecchi M, Plangger K, West D (2021) Supply chain transparency: a bibliometric review and research agenda. Int J Prod Econ 108152
30. Daneshvar Kakhki M, Gargeya VB (2019) Information systems for supply chain management: a systematic literature analysis. Int J Prod Res 57(15–16):5318–5339
31. Dobrescu R, Mocanu S, Chenaru O, Nicolae M, Florea G (2021) Versatile edge gateway for improving manufacturing supply chain management via collaborative networks. Int J Comput Integr Manuf 34(4):407–421
32. Liu Z, Li Z (2020) A blockchain-based framework of cross-border e-commerce supply chain. Int J Inf Manage 52:102059
33. Wang Y, Singgih M, Wang J, Rit M (2019) Making sense of blockchain technology: how will it transform supply chains? Int J Prod Econ 211:221–236
34. Dai B, Nu Y, Xie X, Li J (2021) Interactions of traceability and reliability optimization in a competitive supply chain with product recall. Eur J Oper Res 290(1):116–131
35. Dutta P, Choi TM, Somani S, Butala R (2020) Blockchain technology in supply chain operations: applications, challenges and research opportunities. Transp Res Part E Logistics Transp Rev 142:102067
36. Treiblmaier H (2019) Combining blockchain technology and the physical internet to achieve triple bottom line sustainability: a comprehensive research agenda for modern logistics and supply chain management. Logistics 3(1):10
37. Nasir MH, Arshad J, Khan MM, Fatima M, Salah K, Jayaraman R (2022) Scalable blockchains—a systematic review. Futur Gener Comput Syst 126:136–162
38. Di Pierro M (2017) What is the blockchain? Comput Sci Eng 19(5):92–95
39. Tian Z, Li M, Qiu M, Sun Y, Su S (2019) Block-DEF: a secure digital evidence framework using blockchain. Inf Sci 491:151–165
40. Barenji RV (2021) A blockchain technology based trust system for cloud manufacturing. J Intell Manuf 1–15
41. Zhu H, Guo Y, Zhang L (2021) An improved convolution Merkle tree-based blockchain electronic medical record secure storage scheme. J Inf Secur Appl 61:102952
42. Lee Y, Rathore S, Park JH, Park JH (2020) A blockchain-based smart home gateway architecture for preventing data forgery. HCIS 10(1):1–14
43. Liang W, Tang M, Long J, Peng X, Xu J, Li KC (2019) A secure fabric blockchain-based data transmission technique for industrial Internet-of-Things. IEEE Trans Ind Inf 15(6):3582–3592
44. Neudecker T, Hartenstein H (2018) Network layer aspects of permissionless blockchains. IEEE Commun Surv Tutorials 21(1):838–857
45. Saurabh S, Dey K (2021) Blockchain technology adoption, architecture, and sustainable agri-food supply chains. J Clean Prod 284:124731
46. Bamakan SMH, Motavali A, Bondarti AB (2020) A survey of blockchain consensus algorithms performance evaluation criteria. Expert Syst Appl 154:113385
47. Sai AR, Buckley J, Fitzgerald B, Le Gear A (2021) Taxonomy of centralization in public blockchain systems: a systematic literature review. Inf Process Manage 58(4):102584
48. Lu Y (2019) The blockchain: state-of-the-art and research challenges. J Ind Inf Integr 15:80–90
49. Almasoud AS, Hussain FK, Hussain OK (2020) Smart contracts for blockchain-based reputation systems: a systematic literature review. J Netw Comput Appl 170:102814
50. Buterin V (2014) A next-generation smart contract and decentralized application platform. White Paper 3(37)
51. Chen T, Wang D (2020) Combined application of blockchain technology in fractional calculus model of supply chain financial system. Chaos, Solitons Fractals 131:109461
52. Rathee T, Singh P (2021) A systematic literature mapping on secure identity management using blockchain technology. J King Saud Univ Comput Inf Sci

53. Nakamoto S, Bitcoin A (2008) A peer-to-peer electronic cash system. Bitcoin. https://bitcoin. org/bitcoin.pdf, 4
54. Zheng Z, Xie S, Dai HN, Chen X, Wang H (2018) Blockchain challenges and opportunities: a survey. Int J Web Grid Serv 14(4):352–375
55. Tang H, Shi Y, Dong P (2019) Public blockchain evaluation using entropy and TOPSIS. Expert Syst Appl 117:204–210
56. Chang Y, Iakovou E, Shi W (2020) Blockchain in global supply chains and cross border trade: a critical synthesis of the state-of-the-art, challenges and opportunities. Int J Prod Res 58(7):2082–2099
57. Barenji RV, Nejad MG (2022) Blockchain applications in UAV-towards aviation 4.0. In: Intelligent and fuzzy techniques in aviation 4.0. Springer, Cham, pp 411–430
58. Saadati Z, Zeki CP, Vatankhah Barenji R (2021) On the development of blockchain-based learning management system as a metacognitive tool to support self-regulation learning in online higher education. Interact Learn Environ 1–24

Toward Pharma 4.0 in Drug Discovery

Reza Ebrahimi Hariry, Reza Vatankhah Barenji, and Aydin Azizi

Abstract Pharma 4.0 refers to the employment of cyber-physical technology in the pharmaceutical industry, to operate and monitor the different steps of drug discovery, development, and manufacturing facilities. It can be defined as the digitalization of the pharmaceutical industry from the early drug discovery to post-marketing surveillance by a network of connected devices and enterprises that heavily leverages digital models and ontologies. The Pharma 4.0 might be capable of carrying out multiple scheduled steps facilitated by autonomous systems which allow unmanned operation for extended periods. This chapter highlights recent advances in digital technology applications and will integrate them under Pharma 4.0 philosophy in the drug discovery and development stage. The advancement in this way might affect the speed the way medicines are developed by employing high power of data analytics that may make sure the best drugs are brought to market with high quality and less time.

1 Introduction

Biological systems are intricate information-processing machines during development and disease. The discovery and development of new drugs are two of the most critical translational scientific efforts that benefit human health and well-being. It is

R. E. Hariry (✉)
Department of Pharmacology and Toxicology, Ankara University, Ankara, Turkey
e-mail: hariry@ankara.edu.tr; rehariry@gmail.com

Smart Engineering and Health Research Group, Hacettepe University, Ankara, Turkey

R. V. Barenji
Department of Engineering, School of Science and Technology, Nottingham Trent University, Nottingham NG11 8NS, UK

A. Azizi
School of Engineering, Computing and Mathematics, Oxford Brookes University, Wheatley Campus, Oxford OX33 1HX, UK

© The Author(s), under exclusive license to Springer Nature Singapore Pte Ltd. 2023
A. Azizi and R. V. Barenji (eds.), *Industry 4.0*, Emerging Trends in Mechatronics,
https://doi.org/10.1007/978-981-19-2012-7_10

a lengthy and difficult process that can be broadly divided into three stages: preclinical studies, clinical trials, and regulatory approval. The process of drug discovery tries to develop a chemical that is therapeutically effective in curing and treating disease [1]. This method comprises candidate identification, synthesis, characterization, validation, optimization, screening, and efficacy assays. This stage entails a variety of tasks and testing. Researchers work collaboratively to find and optimize possible leads for a certain target [2]. Essentially, the leads must have the desired impact on a specific biological target associated with a disease in order to treat it. This step involves the use of in silico platforms, biochemical experiments, cell cultures, and numerous animal models. This step is divided into the following sub-processes: Target identification and validation, high-throughput or high-content screening, hit identification, assay development and screening, hit-to-lead (H2L), lead generation and optimization, and in vivo and in vitro assays [3, 4].

Technology has been a significant factor in the advancement of medication discovery. Automation, artificial intelligence, machine learning, nanofluidics, image analysis, big data, software, and test technologies have all contributed significantly to obtaining higher-quality data faster. For example, artificial intelligence (AI) combined with new experimental technology is projected to accelerate, reduce the cost, and increase the effectiveness of the search for novel medications [5]. This chapter discus how drug development can be affected with industry 4.0-related technologies wish to boost the quality of drug product and make time to market short.

2 Drug Discovery

2.1 The Drug Discovery Process

The drug discovery process is a lengthy one, lasting up to 13 years. Typically, only one in every 5000 medications advances to the point of commercial approval. Additionally, only 250 of 5000–10,000 therapeutic candidates advance to preclinical testing. The process of bringing a medication from discovery to commercialization is an expensive one. According to a study conducted by the Tufts Center, the cost of developing a new medicine is projected to be roughly $2.6 billion. Apart from the significant expense required to develop a new treatment, the procedure has become more complicated in recent years. Additionally, post-marketing monitoring and development costs are estimated to be as high as $312 million, bringing the total cost of R&D to three billion dollars per medicine [6].

Typically, the process of early drug discovery begins with a screen for potentially active molecules. These chemicals must have a therapeutic effect on the target ailment, and once identified, safety and efficacy testing can commence. Prior to testing medication candidates in any form on humans, their efficacy and tolerability must be established in cell cultures and animals. If the candidates are determined to be poisonous or harmful, they will be removed from consideration for further development. During this stage, the use of animal models for testing can be minimized in order to adhere to higher ethical standards while also saving time and money [7].

Typically, researchers find new pharmaceuticals as a result of the following:

- New insights into a disease process that enable researchers to create a product capable of halting or reversing the disease's symptoms.
- Numerous testing of chemical compounds to determine whether they have any helpful benefits against a wide variety of disorders.
- Currently available medicines that have unforeseen side effects.
- New technologies such as those that enable the targeted delivery of medical items or the manipulation of genetic material.

Thousands of compounds may be candidates for development as a medicinal therapy at this stage of the process. However, after preliminary testing, only a few chemicals appear to be promising and warrant further investigation.

Once a potential molecule is identified for development, researchers conduct trials to ascertain the following:

- How the compound is absorbed, distributed, metabolized, and eliminated.
- Its potential benefits and action mechanisms.
- The optimal dosage.
- The most effective method of administering the medication (such as by mouth or injection).
- Adverse reactions or side effects, which are frequently referred to as toxicity.
- How it impacts different groups of individuals differently (e.g., by gender, race, or ethnic origin).
- Interactions with other medications and therapies.
- Its efficacy in comparison with comparable medications.

Several critical "steps" are completed during the drug discovery process. Academic and industry scientists collaborate to find druggable targets for a particular disease and to develop and optimize drug molecules capable of eliciting an impact on a specific biological target implicated in the disease—with the goal of curing it. This stage of the process is carried out in the laboratory using in vitro and animal models [8].

2.1.1 Target Identification

One of the most critical aspects of developing a successful medicine is having a thorough understanding of the disease's pathophysiology. It is necessary to identify the target for a certain disease, which is typically a molecule involved in gene regulation or intracellular signalings, such as a nucleic acid sequence or protein. To determine which target to pursue, one must first establish that the molecule is "druggable"—that its activity can be altered by an external chemical. This requires an assessment of cellular and genetic targets, genomic, proteomic, and metabolomic analyzes, as well as bioinformatic predictions. When a therapeutic molecule, referred to as a "hit," may affect the biological activity of a relevant biological target, the target is said to be "druggable" [9].

2.1.2 Target Validation

Potential targets can be identified by utilizing publicly available data sets and conducting a literature search. After identifying a target, researchers evaluate its feasibility for drug development before initiating the screening process to identify hits. After identifying a prospective target, researchers must demonstrate that it contributes to the progression of a particular disease and that its activity can be managed. Conducting meticulous and precise target validation trials is critical for the subsequent stages of drug development to be successful. Validation strategies span from in vitro tools to full animal models to patient-based manipulation of a targeted target. While each strategy is legitimate in and of itself, a multi-validation approach considerably increases confidence in the observed conclusion. Antisense technology, transgenic animals, tissue-restricted and/or inducible knockouts, monoclonal antibodies, and chemical genomics are all examples of effective techniques used in the discovery and confirmation of targets [10].

2.1.3 High-Throughput Screening

High-throughput screening (HTS) and high-content screening (HCS) are frequently the most critical steps in the process of drug discovery. HTS is a drug discovery approach that automates the evaluation of a large number of chemical and/or biological molecules against a specific biological target. In the pharmaceutical sector, high-throughput screening technologies are widely utilized, utilizing robots and automation to rapidly assess the biological or biochemical activity of a huge number of compounds, typically pharmaceuticals. After identifying and validating a prospective target, the beginning point can be found by screening a large number of molecules for their ability to interact with it. An assay must be designed ad hoc, which means providing information regarding the molecule's effectiveness and selectivity. They expedite target analysis by allowing for cost-effective screening of vast chemical libraries. HTS is a valuable approach for analyzing pharmacological targets, pharmacological profiling of agonists and antagonists for receptors (such as GPCRs) and enzymes, among other things [11].

The fundamental objective of HTS is to find, via compound library screens, candidates that have the required effect on the target, referred to as "hits" or "leads." This is often accomplished through the use of liquid handling equipment, robotics, plate readers as detectors, and instrumentation control and data processing software.

2.1.4 Lead (Hit) Identification

Identification of a lead or hit is the process of identifying or developing a chemical that has the potential to interact with a previously identified target. Numerous methods can be employed to identify a target molecule. The term "hit" refers to a molecule that must connect with the aforementioned target in order to exert the desired therapeutic effect. Researchers can run screening trials to uncover natural products that could be repurposed as medicines. High-content screening, phenotypic screening, fragment-based screening, structure-based screening, and virtual screening are just a few of the numerous ways utilized to identify hits. Alternatively, synthetic chemicals can be created that are selective for the predicted target while being non-toxic to other biological processes. Along with determining the drug's mechanism of action, preliminary safety testing is performed in cell culture. Additionally, the pharmacokinetics and pharmacodynamics of the medicine are assessed—that is, how it is digested and how it affects various biological functions. Additionally, molecules are being discovered from molecular libraries by combinatorial chemistry, high-throughput screening, and virtual screening [12].

2.1.5 Hit-to-Lead (H2L)

In a screening experiment, a hit compound is a molecule that exhibits the desired type of activity. By refining the screening criteria, lead compounds are picked from a group of hits, allowing for the selection of the most promising molecules for further development. In other terms, a hit is a term that refers to a chemical substance that exerts a desired therapeutic effect on a previously identified target molecule. Similarly, a lead is a byproduct of the screening process that can be employed in later stages [13].

The hit-to-lead (H2L) stage is critical in the early stages of drug discovery. The H2L's primary objective is to identify appropriate leads for progression along the pathway to a clinically active medication. The first compounds are refined utilizing a variety of screening strategies, including high-throughput screening, affinity selection of vast chemical libraries, fragment-based procedures, and target-focused libraries. Using in vitro and computer-based methodologies, the screening procedure seeks to decrease the number of these compounds to a manageable number of more qualified leads. The qualities of the lead compound must be sufficient to evaluate its efficacy in any in vivo model [14].

2.1.6 Lead Optimization

Once a product (or combination of compounds) is found, it must be optimized for efficacy and safety. Synthetic molecules can be designed in such a way that they are less likely to interact with molecules other than the target. Additionally, using two-

and three-dimensional cell culture platforms, the appropriate dosage and method of administration (oral, injectable) are evaluated.

Additionally, this stage comprises safety testing prior to the subsequent preclinical research stage's introduction into several in vivo animal models. While animal models such as mice and rats can be employed at this point, some safety testing are undertaken in vitro first [15].

Currently, an iterative cycle of structure–activity and in silico investigations is performed in conjunction with cellular functional assays to improve the functional features of freshly synthesized therapeutic candidates.

At the conclusion of the early stages of drug discovery, the resulting candidate compounds must be evaluated in conditions similar to those found in living cells.

3   Beyond Traditional Applications

The primary objective of drug discovery research is to uncover medications that have therapeutic effects on the body—in other words, that can aid in the prevention or treatment of a specific condition. Although there are several different types of drugs, many are small chemically synthesized molecules that can specifically bind to a target molecule—usually a protein—involved in a disease. Traditionally, researchers have searched enormous libraries of compounds for molecules with the potential to become drugs. They then subject this to multiple rounds of testing in order to refine it into a potential compound. Recent years have seen an increase in the use of more logical structure-based drug design methodologies. These bypass the earliest screening stages but still require chemists to design, synthesize, and evaluate a large number of molecules. Due to the fact that it is generally uncertain which chemical structures will have both the required biological effects and the qualities necessary to develop into an effective medicine, the process of refining a promising compound into a drug candidate can be both costly and time consuming. Additionally, even if a new medication candidate demonstrates promise in laboratory testing, it may fail in clinical trials. Indeed, less than 10% of therapeutic proposals get to Phase II studies. Given this, it is unsurprising that specialists are increasingly looking to the unmatched data processing capabilities of emerging technologies such as artificial intelligence systems to speed and lower the cost of drug discovery. According to market research firm Bekryl, artificial intelligence has the potential to save the pharmaceutical industry over US$70 billion by 2028. The sheer breadth of the libraries used to screen for novel medication candidates makes it nearly impossible for individual researchers to review everything—which is where AI and machine learning come in handy. These sophisticated tools enable researchers to mine massive databases for hidden insights [16]. Numerous advantages accrue from this action:

- Predicting the attributes of a possible chemical, ensuring that only compounds with desirable qualities are chosen for synthesis—saving time and money by avoiding labor on ineffective compounds.
- Generating suggestions for completely novel compounds, where the "created" molecule is anticipated to possess all of the desired qualities for success—potentially accelerating the development of effective new medications significantly.
- Eliminating the need for repetitive operations such as histology image analysis—resulting in hundreds of person-hours saved in the laboratory.

These are only a few of the potential benefits when considering the early stages of the drug discovery process.

3.1 AI and Machine Learning in Drug Discovery

In recent years, the area of artificial intelligence (AI) has shifted away from purely theoretical studies and toward practical applications. Cost reduction and project acceleration are critical issues for all pharmaceutical companies. Artificial intelligence is increasingly ingrained in the early stages of drug discovery and is expected to drive biomarker discovery and medication design, hence facilitating precision medicine. Artificial intelligence-based methods are increasingly being employed at various phases of the process to boost efficiency and productivity. These include real-time image-based cell sorting and classification, quantum mechanics (QM) calculations of compound properties, computer-aided organic synthesis, designing new compounds, devising assays, and predicting the three-dimensional structures of target proteins, to name a few. Artificial intelligence pioneered the in silico creation of vaccines, proteins, and tiny molecules, paving the way for completely individualized medicine. In general, these processes are time consuming and may be automated and streamlined to significantly speed up the R&D drug development process with the help of AI [17].

Traditional computational methods handle information using manually programmed logical processes and are used to optimize activities that are too complicated or time demanding for human intellect, such as performing a linear regression on a collection of data. On the other hand, AI is connected with problems in which it is extremely difficult to express the answer to a problem using hand-crafted rules. Artificial intelligence (AI) has the potential to significantly increase the likelihood of uncovering novel therapeutic candidates that are commercially viable. For example, AI can aid in structure-based drug discovery by predicting the three-dimensional protein structure based on the chemical environment of the target protein site, thereby aiding in the prediction of a compound's effect on the target as well as safety considerations prior to its synthesis or production. The recognition of photos containing different items or features has been extremely successful using AI technology [18]. Madhukar et al. created BANDIT, a Bayesian machine learning technique for predicting drug binding targets that incorporates numerous data kinds.

BANDIT achieved a 90% accuracy on 2000 + small molecules by integrating public data [19]. Olayan et al. established an approach that increases the accuracy of drug-target interaction prediction by utilizing a heterogeneous graph containing known DTIs with multiple similarities between medications and numerous similarities between target proteins [20].

3.1.1 Machine Learning

Machine learning is a branch of science that studies how machines acquire knowledge via experience. Supervised, unsupervised, and semi-supervised machine learning are the three broad categories of machine learning approaches. Supervised learning is a task-oriented process. The sample data points have been labeled, and the model has been trained on known samples to predict the labels of the test samples. Unsupervised learning, which is data-driven, seeks to discover hidden structures by learning the patterns in the data. Semi-supervised learning begins with labeled data points and seeks to predict the labels of unknown samples using the labeled data. Support vector machine (SVM), random forest (RF), Naive Bayes (NB), K-nearest neighbor (KNN), gradient boosting trees (GBT), Gaussian processes (GP), and extreme gradient boosting are some of the main machine learning models used in binding affinity prediction (XGBoost) [21]. Boopathi et al. identified anticancer peptides using an SVM meta-predictive model. Meng et al. suggested AOP-SVM machine learning techniques for classifying antioxidant proteins based on sequence characteristics [22]. Houssein et al. introduced two classification techniques, such as HHO-SVM and HHO-kNN, that combine a novel metaheuristic algorithm called Harris Hawks optimization (HHO) with support vector machines (SVMs) and K-nearest neighbors (kNN) for chemical descriptor selection and chemical compound activity [23]. Batool et al. described a support vector machine (SVM)-based predictor (PVP-SVM) for predicting bacteriophage virion proteins in order to aid in the development of new antibacterial treatments. Manavalan et al. developed a support vector machine classifier to predict prognosis and to identify a subset of patients with gastric cancer who may benefit from adjuvant chemotherapy [24].

3.1.2 Deep Learning

Deep learning is a class of machine learning approach that uses artificial neural networks (ANNs) with many layers of nonlinear processing units for learning data representation [25]. In medicinal chemistry, neural networks have been used to classify compounds, perform QSAR analyzes, and identify therapeutic targets [26]. Deep neural networks (DNNs) are neural networks that are completely linked and include numerous hidden layers. Each concealed layer is composed of several nonlinear processing units. DNNs extract characteristics in hierarchical levels automatically by utilizing numerous neurons in many layers. Stacked autoencoder (SAE), variational autoencoder (VAE), deep convolutional neural networks (DCNNs), graph convolutional networks (GCNs), generative adversarial networks (GANs), and recurrent long short-term memory (RLSM) are some of the DNNs utilized in binding affinity prediction (LSTM) [27].

More recently, advancements in new machine learning algorithms, such as deep learning (DL), that generate powerful models from data, and the demonstrated success of these techniques in a number of public contests have contributed significantly to the exponential growth of machine learning applications within pharmaceutical companies over the last two years. When combined with endlessly scalable storage, the significant rise in the types and quantities of data sets that can be used to train machine learning algorithms has enabled pharmaceutical businesses to access and organize a vast amount of data. Images, textual data, biometrics, and other data from wearables, assay data, and high-dimensional omics data are all examples of data kinds [28].

3.2   Chemoinformatics

Cheminformatics (alternatively spelled chemoinformatics) is the application of physical chemistry theory to computer and information science techniques—so-called "in silico" approaches employed in the drug discovery process. Chemoinformatics is the integration of these information resources with the goal of transforming data into information and information into knowledge in order to make better decisions faster in the area of drug lead identification and optimization. Chemoinformatics is a broad topic of study that encompasses a huge number of computational approaches with overlapping applications in a variety of different branches of chemistry [29].

Compound selection, virtual library development, virtual high-throughput screening, data mining, QSAR, and in silico ADMET prediction are all examples of cheminformatics applications in drug discovery [30].

3.2.1   Virtual Library Generation

The computer development of virtual chemical libraries for use in various virtual screening procedures simplifies the process of discovering new hit compounds. As a result, numerous academics are generating new de novo chemical libraries and "make-on-demand" libraries using a variety of in silico methodologies. To save time and money, virtual screening can be used to limit down the number of compounds to examine as prospective lead candidates prior to production and experimental testing [31].

Although the pharmaceutical industry has a long history of utilizing novel methodologies and techniques, there is an urgent need for new ways that might improve and optimize the pipeline of drug discovery and development procedures. Despite the fact that there is no one-size-fits-all solution to this innovation demand, there has been a surge of interest in the usage of integrative systems for this purpose in recent years [32].

The current early stage of drug discovery is plagued by the following issues:

- The used devices are insufficiently intelligent and are not connected to one another.
- All players in the drug discovery process are unable to collaborate effectively.
- All decisions are made by humans, and even well-organized automations are incapable of being proficient.
- Regulatory organizations are not involved in the cycles and experimentation.
- Collaborative efforts between academics, industry, and government are insufficient.
- This cycle continues to be extremely lengthy, costly, and time demanding.

The life cycle of future drug discovery and development will be performed by the following innovations:

- All resources will be connected to one another and to a cloud computation unit where AI technologies will be available and ready to act. The resources will be lot more intelligent.
- Resources will collaborate with one another, obviating the need for repeated processes.
- Using cloud computing to do deeper computational chemistry by selecting relevant chemicals that interact more efficiently with biological targets.
- Tracking and tracing buried products will be significantly easier.
- Having a backup data set and access to irrelevant data at all stages of the experiment.
- Appearance of "as a services" philosophy in drug development life cycle.
- Intelligent drug discovery design.
- Regulatory agencies monitoring and supervising all phases of the drug discovery life cycle.

Pharma 4.0 which is the analog of Industry 4.0 in pharmaceutical science and industry is a framework for adapting the mentioned digital strategies to the unique contexts to incorporate advanced digital elements and enablers [33].

4   Pharma 4.0 for Drug Discovery

Industry 4.0 is defined by integrated, self-organizing, and autonomous production systems. The digital technologies enable higher automation, predictive maintenance, self-optimization of process improvements, and most importantly, a hitherto unattainable degree of efficiency and responsiveness to customers. The principles and technologies associated with Industry 4.0 can be applied to all types of industrial businesses [34]. These digital elements and enablers of Industry 4.0 are as follows:

- Using cyber-physical systems throughout the product life cycle

- Cyber is an intangible asset that should be leveraged to increase computation power.
- Physic is palpable and is utilized to compile data and execute commendations.
- It is a system because the physical world is linked to the virtual world via the Internet via the usage of Internet of things (IoT) technology.

The primary goal of implementing cyber-physical systems is to increase the intelligence of the resource (e.g., gadget, human) and the degree of integration between the resources. Utilizing advanced IoT devices in smart factories results in increased productivity and quality. Substituting AI-powered visual insights for manual inspection business models decreases manufacturing errors and saves money and time. Quality control staff can easily set up a smartphone connected to the cloud to monitor industrial operations from nearly anywhere with no expenditure. Manufacturers can spot mistakes instantly using machine learning algorithms, rather than at a later time when repair work is more expensive. Industry 4.0 has been characterized as "a term that refers to the present trend of automation and data exchange in manufacturing technologies, such as cyber-physical systems, the Internet of things, cloud computing, and cognitive computing, as well as the development of the smart factory." Industry 4.0 is fundamentally altering how businesses make, improve, and distribute their products. Manufacturers are integrating new technologies, such as the Internet of things (IoT), cloud computing, and analytics, as well as artificial intelligence and machine learning, into their manufacturing facilities and processes. From sensors and software to data-driven services across the manufacturing process, Industry 4.0 solutions provide new levels of transparency, quality, safety, and efficiency. In the pharmaceutical sector, particularly, sophisticated processes must be followed, as well as severe regulatory standards. Nobody can accomplish this on their own [35].

Industry 4.0 for medicines will require new ideas to overcome the inertia of current industrial and non-industrial infrastructure, operations, and regulation. In addition, in pharmaceutical research and industry, digital transformation as one of the main factors enables the pharmaceutical business to implement Pharma 4.0 as the pharmaceutical analog of Industry 4.0 pillars and concepts.

Pharma 4.0 refers to the integration of cyber-physical systems in the pharmaceutical sector, in which networked embedded computers operate and monitor drug discovery, development and manufacturing facilities, typically via feedback loops, and physical processes receive feedback from cloud-based computation units. Pharma 4.0 can be characterized as the digitalization of the pharmaceutical industry from the early drug discovery, manufacturing (including planning), and delivery operations perspectives by a network of connected enterprises that heavily leverages digital models and ontologies. CPS applications in the pharmaceutical sector have the potential to dwarf the twenty-first century's pharma 4.0 revolution. They include high-confidence medical equipment and systems, medication design and development, on-demand pharmaceutical manufacturing, 3D-printed pharmaceutical items, logistic 4.0 for drug distribution, and green west management. Control and monitoring of pharmaceutical manufacturing systems could benefit significantly from enhanced embedded intelligence, which would increase quality, safety, and efficiency. Networked autonomous facilities have the potential to significantly improve the effectiveness of our drug development efforts and to provide significantly more effective defect recovery procedures. Networked building control systems (such as cyber-physical embedded process analytical techniques (PAT)) have the potential to dramatically increase the system's intelligence and product quality while reducing our reliance on human judgment. It is easy to foresee new possibilities for drug discovery, development and manufacture, such as "collaborative design and development," where timing precision and quality concerns are minimized. In monitoring and regulatory organizations, distributed agreement about available bandwidth and distributed control technologies could be tremendously beneficial. Precision timing has the potential to significantly alter consumer networks. Distributed real-time quality controllers with sensors and actuators have the potential to fundamentally alter the character of on-line RTRT. Any of these uses would have a massive economic impact. However, pharmaceutical industries may obstruct development toward these uses unnecessarily today. There are significant obstacles, particularly because pharmaceutical discovery and manufacturing systems must adhere to stringent instructions from regulatory organizations in order to meet safety and dependability criteria. Additionally, the tools used to design, monitor, and manage the drug production system (e.g., PAT, QbD, RTRT) are fundamentally distinct from those used in other manufacturing industries. Figure 1 shows the system integration in Pharma 4.0 framework in drug discovery process [36].

The absence of appropriate temporal semantics and concurrency models in computing, along with today's "best-effort" networking technologies, makes predictable and dependable real-time performance challenging. Software component technologies, such as object-oriented design and service-oriented architectures,

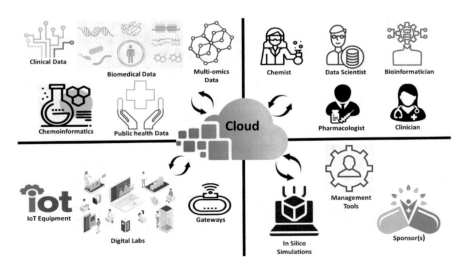

Fig. 1 System integration in Pharma 4.0 framework in drug discovery

are based on abstractions that are more closely associated with software than with physical systems. Numerous applications will be impossible to implement without significant changes to the fundamental abstractions. The digital elements that are included in Pharma 4.0 framework are followings:

Cloud-Based Autonomous Systems

Cloud computing, when paired with the continuous effect of other breakthrough technologies, created an unparalleled opportunity for drug research to benefit from low-cost high-performance computing. Figure 2 shows the combined cloud-based Pharma 4.0 framework in drug discovery. This combination has elevated the cloud to the most promising venue for drug discovery and enables the combination of novel methodologies to achieve success at a higher level of drug development. Cloud computing provides a solution by enabling more complex computational chemistry combined with quick synthesis and optimization, which enables enterprises to find molecules that are more amenable to development from the start. Selecting compounds with a higher probability of success, therefore, functions as a lever for cost reduction, which is necessary if personalized treatment is to be achieved. It contributes to efficiency gains by selecting chemicals that are more compatible with biological targets and have favorable "drug-like" qualities. Advanced experimental methods employed in HTS at various stages of drug discovery (target identification, target validation, lead identification, and lead validation) can generate data in the terabyte range. As a result, a critical requirement exists to store, organize, mine, and analyze this data in order to find informational tags. This requirement puts computer scientists in a position to provide the necessary hardware and software infrastructure, while also managing the varying degrees of expected processing power. As a result, the value proposition of "on-demand hardware" and "software as a service (SAAS)" delivery mechanisms cannot be rejected. This on-demand computing, dubbed cloud computing, is changing drug discovery research. Additionally, cloud computing's integration with parallel computing is undoubtedly growing its footprint inside the health sciences sector. Cloud computing's speed, efficiency, and cost-effectiveness have elevated it to an almost have tool for researchers, offering them tremendous flexibility and allowing them to focus on the "what" of science rather than the "how" [37].

Cloud computing aids in the design of compounds for difficult-to-drug targets by offering ample processing capacity that enables the use of more time consuming yet atomically precise in silico design procedures. Previously, difficult-to-drug targets are showing promise with the addition of computational capacity via cloud computing [38].

Quantum molecular design develops therapeutic candidates and identifies companion biomarkers through cloud-based molecular simulation [39]. Quantum molecular design advances traditional methods by providing a novel method for discovering new molecules through the use of a proprietary AI/"big data" approach, accurate prediction of the binding affinity between a protein and a small molecule, and filtering molecules for desirable chemical properties for drugs. Recent advancements in fields such as microfluidics-assisted chemical synthesis and biological testing, as

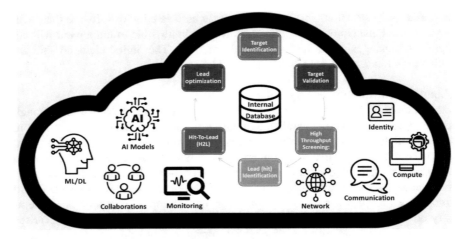

Fig. 2 Cloud-based Pharma 4.0 framework in drug discovery

well as artificial intelligence systems that refine design hypotheses through feedback analysis, are laying the groundwork for the automation of portions of this process [40].

Internet of Things

The Internet of things, or IoT, is a network of interconnected computing devices, mechanical and digital machinery, items, animals, and people with unique identifiers (UIDs) and the ability to send data over a network without requiring human-to-human or human-to-computer interaction. It is a network of physical objects— "things"—embedded with sensors, software, and other technologies with the purpose of connecting to and exchanging data with other devices and systems via the Internet. These devices range in complexity from simple household items to highly sophisticated industrial machines [41].

The Internet of things in healthcare provides interoperability, artificial intelligence machine-to-machine connection, information exchange, and data migration, all of which contribute to the effectiveness of healthcare service delivery. The Internet of things (IoT) has the potential to transform pharmaceutical manufacturing by revolutionizing procedures ranging from drug research to remote patient access and monitoring. By successfully connecting network, equipment, and systems across the plant, IoT technologies promote standardization within the pharmaceutical business. Additionally, by leveraging IoT, pharmaceutical organizations can receive real-time data and visibility into activities throughout the processing chain. Through the use of IoT technology, organizations can achieve uniformity, reduced cycle times, and data integrity. Similarly to AI and machine learning, the Internet of things (IoT) presents a slew of new potential for the pharmaceutical business. IoT sensors and trackers, in particular, can optimize conditions for handling biomaterials and chemicals, assuring flawless equipment operation, and even assisting in the detection of medication fraud [42].

Big Data Analytics

Big data is a topic of study that focuses on methods for analyzing, extracting information from, or otherwise dealing with data volumes that are too massive or complicated for typical data processing application software to handle. While data with a large number of fields (columns) provides greater statistical power, data with a higher level of complexity (more attributes or columns) may result in a higher rate of false discovery. The issues associated with big data analysis include data collection, data storage, data analysis, search, sharing, transfer, visualization, querying, and updating, as well as information privacy and data source. Initially, big data was associated with three fundamental concepts: volume, variety, and velocity. Because the analysis of huge data offers sampling issues, it was previously limited to observations and sample. As a result, big data frequently comprises data that is larger beyond the capacity of standard software to process in a reasonable amount of time and money [43].

Big data is defined by three characteristics: volume, diversity, and velocity. These features, taken together, describe "big data." They have necessitated the development of a new class of capabilities to complement the way things are now done in order to provide a clearer view and control over our existing knowledge domains, as well as the capacity to act on them. Big data has changed the way we manage, analyze, and leverage data across industries. Healthcare is one of the most noteworthy domains where data analytics is having a significant impact. In healthcare, the term "big data" refers to the huge amounts of information generated by the adoption of digital technology that collect patient records and aid in the management of hospitals that are otherwise too large and complex for traditional technologies. In essence, big-style data refers to the massive amounts of information generated as a result of the digitization of everything and consolidated and analyzed using specialized technologies [44].

Envisioning big data's future role in digital healthcare requires balancing the benefits of improved patient outcomes against the potential pitfalls of increased physician burnout as a result of poor implementation resulting in added complexity. Pharmaceutical businesses have traditionally depended on empirical data to uncover patterns, test hypotheses, and gain a better understanding of treatment efficacy. Data analytics is merely the latest step in a process that has been going on for hundreds of years: increasing human access to knowledge and data. Since the introduction of the moveable type printing press in the fifteenth century, the creation of human knowledge and data has arguably accelerated. This single breakthrough-enabled unprecedented information dissemination—scientists could now simply discuss the results of research conducted in one country with scientists in other countries. Numerous academic institutions and industrial research institutes collaborate globally to undertake one or more aspects of the drug design and development process [45, 46].

Pharma 4.0 may offer a one-of-a-kind platform for integrating all of these stakeholders in real time. This concurrency has a beneficial effect on cycle time and costs, as well as avoiding several misconceptions and prejudices (Fig. 1). In the Pharma 4.0 age, the cloud will advise daily work schedules for operators and research centers. This will eliminate unneeded experiments, which are a significant source of waste.

The time, cost, and quality of developing a novel molecule are critical enablers for pharmaceutical businesses worldwide. They are constantly on the lookout for new ways to manage and optimize these enablers.

5 Conclusion

Pharma 4.0 is a trend for adapting digital tactics to the specific circumstances of various pharmaceutical manufacturing processes. Pharma 4.0 is defined by the incorporation of modern digital elements and enablers such as cloud computing, artificial intelligence and machine learning tools, big data, IoT technology, simulation, and virtualization in a multi-agent-based integrated autonomous system. By facilitating early communication between all of their branches and R&D departments, academic investigators, and regulatory bodies, this cyber-physical system enables pharmaceutical corporations to achieve the aforementioned aims. Additionally, it integrates the system with other businesses, health technology assessment organizations, and investors in accordance with their own privacy rules.

By implementing real-time supervision in the Pharma 4.0 era, regulatory agencies will be able to more closely monitor each aspect of the drug development process. This may require broadening the process's scientific foundations, managing all safety and efficacy assessments, monitoring all phases, and improving preclinical and clinical registrations.

Pharma 4.0 contributes to the drug development process in the following ways:

- Decrease the time required for medicinal development.
- Develop more tolerable novel chemicals for the purpose of target discovery.
- More accurately predict the toxicity and adverse consequences of substances.
- Conduct fewer but more focused experiments.
- Accelerated progression through the preclinical and clinical phases of drug development.
- Decreased development resources.
- Minimize erroneous judgments and experimenter biases.
- Using AI tools, design logical and intelligent tests.
- Cost reduction of experiments and avoidance of duplications.
- encapsulating all findings in a knowledge model for future investigations.

References

1. Mohs RC, Greig NH (2017) Drug discovery and development: role of basic biological research. Alzheimer's Dement Transl Res Clin Interventions 3(4):651–657
2. Dickson M, Gagnon JP (2009) The cost of new drug discovery and development. Discov Med 4(22):172–179
3. Dai L, Li Z, Chen D, Jia L, Guo J, Zhao T, Nordlund P (2020) Target identification and validation of natural products with label-free methodology: a critical review from 2005 to 2020. Pharmacol Ther 216:107690
4. Holdgate G, Embrey K, Milbradt A, Davies G (2019) Biophysical methods in early drug discovery. ADMET DMPK 7(4):222–241
5. Hill RG, Richards D (2021) Drug discovery and development E-book: technology in transition. Elsevier Health Sciences
6. DiMasi JA, Grabowski HG, Hansen RW (2015) The cost of drug development. N Engl J Med 372(20):1972
7. Hughes JP, Rees S, Kalindjian SB, Philpott KL (2011) Principles of early drug discovery. Br J Pharmacol 162(6):1239–1249
8. Fotis C, Antoranz A, Hatziavramidis D, Sakellaropoulos T, Alexopoulos LG (2018) Network-based technologies for early drug discovery. Drug Discovery Today 23(3):626–635
9. Schenone M, Dančík V, Wagner BK, Clemons PA (2013) Target identification and mechanism of action in chemical biology and drug discovery. Nat Chem Biol 9(4):232–240
10. Blake RA (2007) Target validation in drug discovery. In: High content screening. Humana Press, pp 367–377
11. Szymański P, Markowicz M, Mikiciuk-Olasik E (2012) Adaptation of high-throughput screening in drug discovery—toxicological screening tests. Int J Mol Sci 13(1):427–452
12. Agamah FE, Mazandu GK, Hassan R, Bope CD, Thomford NE, Ghansah A, Chimusa ER (2020) Computational/in silico methods in drug target and lead prediction. Brief Bioinform 21(5):1663–1675
13. Hoffer L, Muller C, Roche P, Morelli X (2018) Chemistry-driven hit-to-lead optimization guided by structure-based approaches. Mol Inf 37(9–10):1800059
14. Heifetz A, Southey M, Morao I, Townsend-Nicholson A, Bodkin MJ (2018) Computational methods used in hit-to-lead and lead optimization stages of structure-based drug discovery. In: Computational methods for GPCR drug discovery. Humana Press, New York, NY, pp 375–394
15. Joseph-McCarthy D, Baber JC, Feyfant E, Thompson DC, Humblet C (2007) Lead optimization via high-throughput molecular docking. Curr Opin Drug Discov Devel 10(3):264–274
16. Paul D, Sanap G, Shenoy S, Kalyane D, Kalia K, Tekade RK (2021) Artificial intelligence in drug discovery and development. Drug Discovery Today 26(1):80
17. Smith JS, Roitberg AE, Isayev O (2018) Transforming computational drug discovery with machine learning and AI. ACS Med Chem Lett 9(11):1065–1069
18. Fleming N (2018) How artificial intelligence is changing drug discovery. Nature 557(7706):S55–S57
19. Madhukar NS, Khade PK, Huang L, Gayvert K, Galletti G, Stogniew M, Allen JE, Giannakakou P, Elemento O (2019) A Bayesian machine learning approach for drug target identification using diverse data types. Nat Commun 10(1):1–14
20. Olayan RS, Ashoor H, Bajic VB (2018) DDR: efficient computational method to predict drug–target interactions using graph mining and machine learning approaches. Bioinformatics 34(7):1164–1173
21. Vamathevan J, Clark D, Czodrowski P, Dunham I, Ferran E, Lee G, Li B, Madabhushi A, Shah P, Spitzer M, Zhao S (2019) Applications of machine learning in drug discovery and development. Nature Rev Drug Discovery 18(6):463–477
22. Boopathi V, Subramaniyam S, Malik A, Lee G, Manavalan B, Yang DC (2019) mACPpred: a support vector machine-based meta-predictor for identification of anticancer peptides. Int J Mol Sci 20(8):1964

23. Houssein EH, Hosney ME, Oliva D, Mohamed WM, Hassaballah M (2020) A novel hybrid Harris hawks optimization and support vector machines for drug design and discovery. Comput Chem Eng 133:106656
24. Manavalan B, Shin TH, Lee G (2018) PVP-SVM: sequence-based prediction of phage virion proteins using a support vector machine. Front Microbiol 9:476
25. Chen H, Engkvist O, Wang Y, Olivecrona M, Blaschke T (2018) The rise of deep learning in drug discovery. Drug Discovery Today 23(6):1241–1250
26. Gawehn E, Hiss JA, Schneider G (2016) Deep learning in drug discovery. Mol Inf 35(1):3–14
27. Lavecchia A (2019) Deep learning in drug discovery: opportunities, challenges and future prospects. Drug Discovery Today 24(10):2017–2032
28. Zhang L, Tan J, Han D, Zhu H (2017) From machine learning to deep learning: progress in machine intelligence for rational drug discovery. Drug Discovery Today 22(11):1680–1685
29. Lo YC, Rensi SE, Torng W, Altman RB (2018) Machine learning in chemoinformatics and drug discovery. Drug Discovery Today 23(8):1538–1546
30. Martinez-Mayorga K, Madariaga-Mazon A, Medina-Franco JL, Maggiora G (2020) The impact of chemoinformatics on drug discovery in the pharmaceutical industry. Expert Opin Drug Discov 15(3):293–306
31. van Hilten N, Chevillard F, Kolb P (2019) Virtual compound libraries in computer-assisted drug discovery. J Chem Inf Model 59(2):644–651
32. Izmaylov A, Saraev A, Barinova Z (2021) The development of the domestic pharmaceutical industry in the context of digitalization. In: Current achievements, challenges and digital chances of knowledge based economy. Springer, Cham, pp 181–188
33. Hariry RE, Barenji RV, Paradkar A (2020) From Industry 4.0 to Pharma 4.0. In: Handbook of smart materials, technologies, and devices: applications of Industry 4.0, pp 1–22
34. Frank AG, Dalenogare LS, Ayala NF (2019) Industry 4.0 technologies: implementation patterns in manufacturing companies. Int J Prod Econ 210:15–26
35. Barenji RV, Akdag Y, Yet B, Oner L (2019) Cyber-physical-based PAT (CPbPAT) framework for Pharma 4.0. Int J Pharm 567:118445
36. Hariry RE, Barenji RV, Paradkar A (2021) Towards Pharma 4.0 in clinical trials: a future-orientated perspective. Drug Discovery Today
37. Vatankhah Barenji R (2021) A blockchain technology based trust system for cloud manufacturing. J Intell Manuf 1–15
38. Garg V, Arora S, Gupta C (2011) Cloud computing approaches to accelerate drug discovery value chain. Comb Chem High Throughput Screening 14(10):861–871
39. Li J, Topaloglu RO, Ghosh S (2021) Quantum generative models for small molecule drug discovery. IEEE Trans Quantum Eng 2:1–8
40. Keinan S, Frush EH, Shipman WJ (2018) Leveraging cloud computing for in-silico drug design using the quantum molecular design (QMD) framework. Comput Sci Eng 20(4):66–73
41. Wortmann F, Flüchter K (2015) Internet of things. Bus Inf Syst Eng 57(3):221–224
42. Yeole AS, Kalbande DR (2016) Use of internet of things (IoT) in healthcare: a survey. In: Proceedings of the ACM symposium on women in research 2016, pp 71–76
43. Tsai CW, Lai CF, Chao HC, Vasilakos AV (2015) Big data analytics: a survey. J Big Data 2(1):1–32
44. Sestino A, Prete MI, Piper L, Guido G (2020) Internet of things and big data as enablers for business digitalization strategies. Technovation 98:102173
45. Saranya P, Asha P (2019) Survey on big data analytics in health care. In: 2019 International conference on smart systems and inventive technology (ICSSIT). IEEE, pp 46–51
46. Galetsi P, Katsaliaki K, Kumar S (2020) Big data analytics in health sector: theoretical framework, techniques and prospects. Int J Inf Manage 50:206–216

What Military 4.0 IS: Applications and Challenges

Egemen Akçay, Melis Etim, Mahmut Onur Karaman,
and Merve Turanli Parlaktuna

1 Introduction

We are going to discuss Industry 4.0 applications from a military point of view. And while discussing this topic we are going to examine different articles and topics, such as IoT in Military Applications, Blockchain Technology in Military and an example of it and Logistics 4.0 within two different perspective of the two specific articles.

As military operations are modernized and digitalized, a need for digitalized logistics is arise. Logistic 4.0 transformation is not easy in military as the budget and time is limited. Private—public sector collaboration is one of the solutions for a smooth transformation. First, we are going to discuss the push–pull mechanism of the logistics and transaction between them.

Then, the relationship between success of logistics and winning a battle. In twenty-first century it becomes much important because of the complexity of the military challenges so that experts try to improve new ways to enhance the power of logistics. In battle conditions the necessary elements must have delivered as complete, flexible, sufficient and in time. And this necessity highlights the importance of the logistics network.

For deep diving into tracking and tracing systems, we are going to examine the blockchain technology briefly with an example of supply chain management and after that we are going to try to understand the past of military success and the correlation between success and tactics in battle. In near future, Internet of Things technology is the most widely used Industry 4.0 technologies in real life. It is a technology with simple architecture, low cost and high efficiency. It is user-friendly thanks to its simple interfaces. It has no decision-making responsibility. It stores the information it senses and collects the data and transmits it to the decision maker. For

E. Akçay (✉) · M. Etim · M. O. Karaman · M. T. Parlaktuna
Department of Industrial Engineering, Hacettepe University, Beytepe Campus, 06800 Ankara, Turkey
e-mail: akcay.egemn@gmail.com

© The Author(s), under exclusive license to Springer Nature Singapore Pte Ltd. 2023
A. Azizi and R. V. Barenji (eds.), *Industry 4.0*, Emerging Trends in Mechatronics,
https://doi.org/10.1007/978-981-19-2012-7_11

this reason, it is very useful for military areas where track and tracing systems are used. In this section, the use of IoT technologies in the military field is mentioned. Human supervised and autonomous systems according to its usage fields are pointed out in the article. In addition, IoT applications which are used in the military field are mentioned. On the other hand, military areas cannot be considered only as battlefield. Some IoT applications which are useful for military logistics and health monitoring of soldiers are explained.

2 Application of Industry 4.0 in Military Sector

2.1 Logistic

2.1.1 Military Logistics

NATO defines logistics as "the science of planning and carrying out the movement and maintenance of forces." Military operations could not be carried out and sustained without logistics [1] (Fig. 1).

Nowadays, military operations are affected by digital transformation. While military operations are modernized and digitalized, the need for a new logistics system also arise. Efficiency and optimization are always important for logistics however, today other factors should be taken account due to digital transformation of military

Fig. 1 Military logistic visualization [1]

operations and the rise of IoT technologies utilization. The need of digital transformation in logistics is obvious however the methodology is not defined precisely. Military logistics have different and significant issues to consider. One of the differences is the highly uncertain conditions. While perfect forecasts are possible in other sectors, it is not possible in defense sector. There are difficult questions to answer such as what supplies or services will be needed, where and when they will be needed, or the best way to provide them. Organizational change and strategic support is necessary to provide an easy transition to new logistic 4.0 from traditional one [2].

The new logistics is based on a conceptual shift away from a "push" orientation toward a "pull" orientation. This indicates that the supply type changes from one that is driven by offer to one that is driven by demand. New logistical models are more efficient, robust, sustainable, intelligent, agile, adaptive, flexible, and scalable [3].

The following are the distinctions between traditional and modern logistics:—Competition for service rather than pricing,—Frequent transportation of small lots—Adoption of a just-in-time strategy—the objective is to maintain no inventory.—Establish global logistics networks with integrated distribution facilities;—Outsourcing;—Being more environmentally mindful and concerned about sustainability. To accomplish the aforementioned qualities, a physical Internet idea is presented. The primary objective is to decrease inefficiencies in global transportation and logistics network waste management. The internet term is used metaphorically. The ultimate objective is to establish a logistic network capable of facilitating global information movement over the Internet. The United States of America and the European Union have embraced a vision of physical Internet as the logistics conceptual aim for 2050. Physical Internet implementation will necessitate the use of current digital enabler technologies, such as the Internet of Things, cyber physical systems, collaborative robots, ecommerce, blockchain, virtual reality, augmented reality, and big data analytics. Due to the difficulty of forecasting military demand variations, it is critical to have increased connection and integration between the final and initial linkages of global supply chains, as well as increased capability for data aggregation and automated estimates. A study of the inconsistencies and delays in adapting to new logistics 4.0 of the military and private sector [2].

The study is conducted with the participants of defense sector that accounted for approximately 85% of sales in 2016. The results show that almost all companies that are participated in the study, consider that digital transformation in logistics means opportunities for improvement in the efficiency, quality, and flexibility of processes, which will turn as extra profit. However, none of the companies is proactive for change due to limitations in budgets and also lack of strategic plans. It is concluded that the gap between logistics 4.0 will tend to grow due to lack of proactive adaption of defense sector. The article suggests outsourcing; the transfer of operational activities of military to private sector as a solution. Some companies in private sector proactively adopted to changes in new logistics system. In order to balance between private sector and armed and police forces, public—private cooperation models are suggested. By this way, in addition to involving advantages in the improvement of military and police efficiency, private sector will also benefit by participating bigger and more complex projects. Furthermore, transfer of knowledge both from private sector to public and public to private sector will catalyze the adaptation and implementation of logistics 4.0. In conclusion it is a fact that, military logistics has significance importance for military operations sustainability and continuity. As the digital transformations are realized in every aspect of the life, logistics area also evolve accordingly. Generally, public sector cannot spare time and allocate budget for military logistics digitalization, thus know-how of private sector companies can be utilized to achieve logistic 4.0 goals.

2.1.2 Functional and Relational Determinations of Modern Operational Military Logistics

Overview of Industry 4.0 and Supply Chain Management 4.0

On the basis of this drive, new requirements emerge in order to capitalize on emerging technical potential in the economic and logistical sectors, under the new conditions dictated by today's automated and digital nature.

I believe it is critical to give a synthesis of Industry 4.0 and supply chain management 4.0, particularly in the economic sector.

Industry 4.0's functional consequences are visible in the design, implementation, and current systemic operation of domestic and transnational supply chains. It is critical that clear connections between the Internet of Things, advanced robotics, advanced data management, expanded automation, sensor deployment in any process, and generalization of network operation are established to fully raise performance standards in the efficient operation of each supply chain and meeting customer needs as beneficial effects of Industry 4.0. I give a broad view, a framework for understanding what is discussed in Fig. 2.

In comparison to traditional supply chains, supply chains 4.0 enable continual improvement in the efficiency of products and related services flows from suppliers to beneficiaries while minimizing costs associated with supply, manufacture, and distribution to end consumers [5, 6].

Influences of Industry 4.0 and Supply Chain 4.0 Management on Operational Logistics

Given the integration of operational logistics foundations into all Alliance and combined national combat forces organizations, it is viewed as a distinct, inter-systemic system in terms of innovation and modernization. From this point forward, the supply and resupply of operating forces begins to be integrated with modern technical systems and combat materials, as well as effective and efficient modes of movement and transport, all of which offer adequate capabilities for concealment, self-protection, deployment, and use in complex environments associated with future wars. There is confrontation between opposing forces. Figure 3 illustrates the many

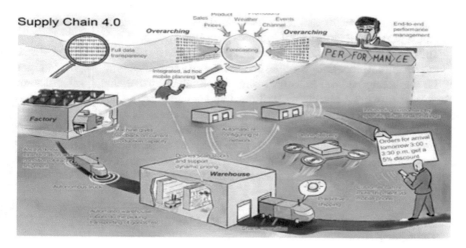

Fig. 2 Synthetic vision on the supply chain management mechanism 4.0 [4]

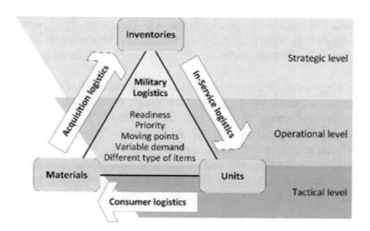

Fig. 3 Relation of the types of military logistics in the process of achieving the logistical support of the joint [5]

forms of military logistics used to supply essential commodities and services to consumers and users inside active military organizations.

To successfully meet the above requirements, military logistics managers and their subordinates in all areas of operational logistics support are responsible for coordinating the acquisition, installation, and use, for the purpose of increasing efficiency, of new technologies already widely used in the digital society, including artificial intelligence, such as "big data; cloud computing; Internet of Things (IoT); autonomous vehicles; robotics; and 3D printing," among others. Through their unique hardware and software components, as well as their functional integration, the aforementioned technologies functionally define contemporary and complicated digital mechanisms that must provide the operational logistics' increased power [7].

Thus, goods, processes, and materials susceptible to quick acquisition and distribution to military operations organizations require the successful implementation of 4.0 marketing and 4.0 logistics inside supply chain management 4.0. All three disciplines provide AI-aware capabilities that are suited to the increasing demands of combat military units based on their functional features and missions.

Military logistics has evolved excellently in the modern era, becoming more complicated. This is because each type of military troops provides particular support to operational organizations, which requires proper information and cyber security (along with other forms of force protection).

2.1.3 Essential Elements of the Modern Funtioning of Operational Logistics

As a result, they build organizations that distribute products, processes, and materials, as well as conduct quick purchasing and military operations, which includes the successful use of 4.0 marketing and 4.0 logistics inside the supply chain management 4.0. Depending on their purpose and goal, each of the three categories provides AI-enabled capabilities that are suited to the increasing demands of combat military formations.

In terms of operational force integration in a future complex conflict environment, the relevant new requirements emphasize the critical links and interactions between technological and economic systems that are particularly innovative and efficient, involving supply chains in a continuous interdependence of the combat force structure in order to fulfill contracts with beneficiary organizations.

Technological, economic, and communication improvements enabled by especially simple Smart-Tech are also transforming military logistics integrated into operational structures that are meant to deploy and act in theaters or regions of operations [8].

Military supply chains (modernized in accordance with SCM 4.0 requirements), in order to be integrated into the mechanism for timely provision of logistical support to operational forces, must enable rapid response, effective and efficient operation, adaptation and augmentation capabilities, and the ability to respond immediately to unforeseen requests imposed by the immediate and sustained changes in the actions of fighting forces [9].

As a result, the unique characteristics of national operational forces, as well as those inside the Alliance's structure, need the establishment of supply chains capable of continually responding to dynamic requests for products and services. Additionally, the timely distribution of replacement parts and aggregates required for the maintenance of technological and weapons systems, which are frequently made by unique producers with no identifiable market, is included (for example: engines; transmissions; housings, etc.) [10].

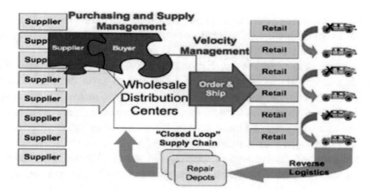

Fig. 4 A model of the supply-delivery chain for the logistical support of the operational forces [9]

Figure 4 illustrates a supply chain model that was employed to assist the logistical support of the United States operating troops.

On this basis, significant technical, economic, digital, and managerial advancements enable commanders of operational forces and their subordinate logistics leaders to plan and assure the efficient and effective use of all logistical resources available to them. This enables them to augment operational capability and promote the successful completion of duties assigned to subordinate troops, regardless of the time, season, or weather conditions in the theater or area of military operations.

I suppose that in the subsequent stage, logistics supervisors and subordinates will carry out scheduled and supervised actions. These operations are designed to facilitate the speedy and secure transportation of material items (products, processes, and materials) across supply chains (civil and/or military).

Among them are the following:

- all types of suppliers (civilian or military) (including suppliers' suppliers);
- various domestic and multinational producers (whether or not in the defense industry);
- centralized distributors (civil distribution centers–zonal and local);
- support groups or modules for the execution of logistic support at the second line of logistic support;
- ANRPS wholesalers and material warehouses or economic operators.

On the basis of the foregoing, I believe that supply chains (civil, military, or mixed) must have the infrastructure and associated modern technologies to enable continuous inflows and outflows (material, financial, informational, and other) to meet the logistical support requirements of operational forces in a particular theater or area of operations.

Although the data and logistical information required to support the profile of operational forces are now fully computerized, capitalized, and properly transmitted, military experts in modern armies report that there are still no efficient and secure technologies (platforms, systems, etc.) that would enable a total and continuous visibility of the existing resources in transit, pre-positioned, on the beneficiary military organizations, as well as of the operations [11]

The disclosed aspects address, on the one hand, possible manifestations of supply chain inadequacy in terms of agility and resilience, and, on the other hand, nonconformities in the continuous, timely, and efficient information of logistics managers and, implicitly, leaders of maneuvre and support structures with current and accurate information, at the appropriate locations and times for operational situations, which may result in incomplete replenishment [12].

The demands of the changing operational needs inherent in today's and tomorrow's military problems dictate that movement and transport constitute a critical component of logistical support, inextricably linked to supply chain management. Thus, in order for the fighting and support forces to fulfill their missions, significant types and quantities of logistic resources must be provided in the indicated locations and at the opportune times, enabling the rapid deployment, integration, and engagement of operational forces in nonlinear offensive and/or defensive, hybrid, and asymmetric actions, in order to fulfill the missions of the Alliance's force structures engaged in joint multinational operations [13].

2.1.4 Aspects of Modernity in the Planning and Realization of Operational Logistics

In modern times, operational logistics planning is a highly important initial step at tactical levels as such planning offers the commander the opportunity to employ the right resources, choose the correct decisive options and achieve a clear victory over opposing forces.

Figure 5 generically highlights "Sense and Respond Logistics" system which connects the source of support source to the point of use within the network. The result is that the traded information is processed in real time through the S&R control node (C2).

In order to satisfy the needs of S&RL authorities, US military professionals emphasize the importance of stimulating innovation, testing, verifying, and validating methods and procedures that support the successful and progressive deployment of the S&RL concept. As a result, this method clearly facilitates the activities of logistics leaders participating in tactical and collaborative logistical support decision-making processes [15].

The United States' 2017 National Security Strategy outlines a series of criteria that emphasize the future focus on expeditionary operations that are adequately backed by required resources within current bases on national territory. This will be accomplished by pre-positioning the stocks essential for the forces that will prepare for and conduct the aforementioned activities, with the assistance of extra forces factored in.

A team of military professionals (active and reserve members from North Carolina) developed and executed a specialized system for military logistics network

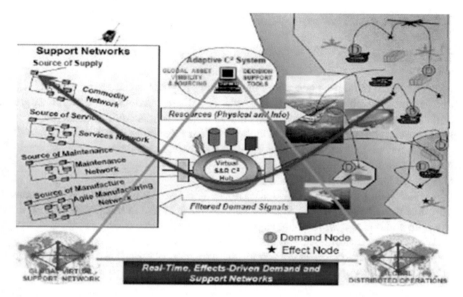

Fig. 5 Generic relevance of the "Sense and Respond Logistics" mechanism [14]

planning based on the aforementioned security strategy (MLNPS). This cutting-edge (planning) mechanism enables user military logisticians to rapidly assess the data and information necessary to identify, develop, evaluate, and compare various courses of action for the successful execution of military expeditionary operations 2020 [11].

Thus, based on the study findings of the aforementioned military specialists, the MLNPS program presently enables the detection of logistical bottlenecks, permitting the short-term highlighting of their implications. It is self-evident that in order to assure the continuity of supply flows necessary to sustain the expeditionary operating organizations' combat activities, their logistics managers must be able to forecast, in advance, any delays downstream of a supply-delivery chain (military, civilian or mixed). For instance, due to functional disruptions, a supplier (contractor) may gradually lower the output established to be provided to consumers for objective reasons (beneficiary maneuvring and support structures). Under these circumstances, logistics leaders will promptly advise commanders of potential procurement choices in order to make timely decisions to ensure enough supply flows essential for the dispatching operational forces to execute their tasks [11, 16].

According to military experts, it is clear that as a result of the functional moderniza-tions, the evolved command and logistics control systems (derived from the endow-ment of the North American military's operational forces) will be capable of ensuring adequate agility, even if enemy forces disrupt communications through hostile cyber-netic and electronic actions. Continuous technical upgrading of logistics data trans-mission systems, from digital processes to analog and functional modes, is critical for the aforementioned aim.

2.2 Track and Tracing Systems

2.2.1 Blockchain

Blockchain technology is becoming increasingly popular in the era of Industry 4.0 and the Information Age. Blockchain is a decentralized and distributed digital ledger or database of records and transactions that helps businesses record transactions and track assets. The use of blockchain technology may also help to prevent counterfeit products from being substituted for genuine products. Transparent processes, traceable and recorded operations are some of the other advantages of using blockchain in Supply Chain Management.

We are going to examine Military Blockchain For Supply Chain Management article of the SB Rahayu, ND Kamarudin, AM Azahari and in this article there is an example and concept work about supply chain traceability.

Westerkamp, Martin, Friedhelm Victor, and Axel Küpper propose a smart contract-based blockchain-based supply chain tracing solution. This concept is based on two main concepts:

1. Introduce digital tokens to represent actual products and preserve a relationship between source and product;
2. Additional features to promote cross-business traceability, such as certifying goods, transferring, splitting, and combining tokens.

Smart contracts are computer programs that enforce laws without requiring the involvement of a third party (Fig. 6).

2.2.2 Military Applicatons of Blockchain Tracking and Tracing Systems

Currently, just a few military defense organizations are looking into how blockchain might help them maximize their capacity and capability in terms of assets, activities, and operations.

The US Department of Defense is experimenting with blockchain technology in defense and security. The Lebanese Armed Forces have expressed interest in using blockchain technology to keep their centralized force running. Military supply chain management is complex and difficult like other sectors that are having security concerns. Tracing defense shipments/contracts, secure government and battlefield messaging, cyber warfare preparedness, preventing data theft, NATO applications, protecting weapons systems, and military additive manufacturing are all conceivable use cases for military blockchain (Fig. 7).

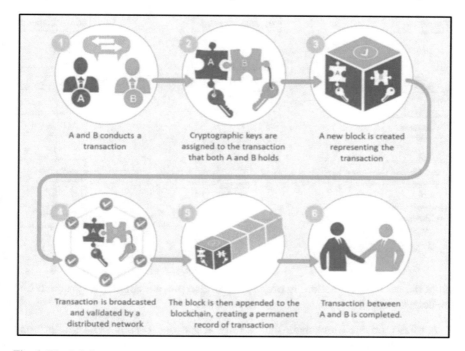

Fig. 6 Blockchain process [17]

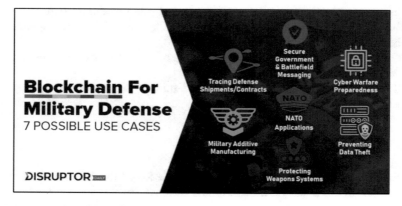

Fig. 7 Seven use cases for military blockchain [18]

A Framework for Military Blockchain

There is an example in the related article that is about Navy Defense Shipment. The purpose is montoring the third-parties all the time. It ensures our system to be sure

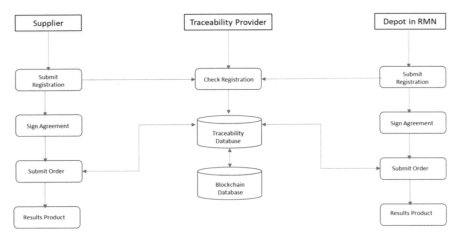

Fig. 8 Military blockchain [18]

about the assets are functioning properly. It is also prevent counterfeit products and the delays (Fig. 8).

- A totally private blockchain that is only accessible to authorized military and replacement parts vendors.
- All spare parts suppliers must register as a traceability-certified organization.
- A Navy Depot, such as West Fleet Supply Depot (WFSD), must register as a business in order to verify the genuineness and validity of spare parts.
- The blockchain database will eventually take the place of the existing system for obtaining spare parts at the Depot.
- A blockchain is a decentralized digital ledger. The ledger validates and archives all general-purpose transactions.
- To ensure safe communication inside the military blockchain, a cryptography technology is used [18].

2.2.3 Internet of Thing in Military

Before talking about why IoT technology is needed in the Military field and its applications in this field, it is necessary to mention the evolutionary development of warfare and previous technological developments that changed the management of warfare. Unfortunately, wars have existed since humanity has existed. Even though their sides and method may change, wars will continue to take place somewhere in the world today and tomorrow. At the beginning of human history, new discoveries were taking place on earth and lands were changing hands frequently between human societies. The first armies were used to chosen from the strongest men of the society, and the armies were equipped with the most technological weapons of that period such as swords, etc. The armies would then meet in a square or on a battlefield and fight until one side gave up or was destroyed. Therefore, the side with more warriors or stronger in terms of people and weapons would win. We can call it the war of the rows and columns. Because when a matrix is considered, all the power of the row player and column player fills the rows and columns of the matrix. Between the two players, the side whose matrix cells have more values wins. This is the most classical period warfare method we know. But later on, a method was developed in which armies with fewer soldiers could defeat armies larger than themselves. This method was theorized by the Prussian war theorist Clausewitz. According to this theory, the final result was based on the principle of gathering combat power at the right place and at the right time. Of course, doing this required better coordination, mobilization, communication, and a more advanced logistics system. Artillery with indirect firepower made the greatest contribution to this revolution.

We said that every new and bright idea is advantageous on the battlefield. Therefore, the concept of war changes. The next revolution that changed the concept of warfare was that tactics. To explain this with an example, we can say that the very important use of cryptography or the effective use of the discipline of operations analysis on the battlefield. These are elements that are outside the fighting personnel but are at least as beneficial as the fighters. It is the product of an intelligence. Everything can be used to be the product of this intelligence. Highways, city infrastructures, computers, scientific disciplines, and many other things can be considered in war tactics. Therefore, this period can be called the age of complex systems. So, this period can be called as the age of complex systems. It also requires greater attention and coordination of the commanding staff.

So how has warfare changed today? First of all, there is a situation that we can define as the twilight zone. In other words, there is uncertainty where the declaration of war is not made openly, but the enemy is worn down without declaring war. Also, the sides of the wars are not clear today. Sometimes there are wars in which small terrorist groups are used, and sometimes there are secret allies. So the allies are also uncertain in most cases. Finally, we can say that today the term battlefield has completely disappeared. Every platform is now a battlefield. First of all, city centers, where the threat is a complete mystery, can now be battlefields. In addition, not only the real world, but also virtual environments are now the battleground. Attacks on a country's internet infrastructure or virtual environments where official records are kept can be an example of this. Or a country may attack another country's economy with various trade sanctions. As a result, wars today occur as asymmetric and hybrid wars with a lot of uncertainty.

All these developments have made it necessary for decision makers to develop a dynamic decision-making mechanism and a network structure to resolve the uncertainty and multilateral war environment. This network structure should be in the form of collecting the information of the tools on the battlefield, ensuring the communication of these tools with each other, then processing the collected data and finally presenting the useful information to the command and control level. It should be noted that surveillance and reconnaissance, command control and communication systems are very important in this system where the network structure is used (Fig. 9).

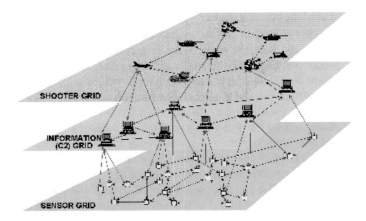

Fig. 9 Network structure

Internet of Things technology is one of today's technologies that will best meet the needs of the concept of today's war, where such a network structure is needed. For this reason, it is considered that the Internet of Things will be very useful in the military field and will be used more frequently by countries in the coming years. On the other hand mostly, military fields have been the pioneers in technological developments. The reason for this is that countries invest a lot in the field of defense and a new technological application in the military field provides a great advantage against the enemy. For this reason, military examples of the Internet of Things will be applied very often in the near future.

IoTs may help military with five ways. These are;

1. Providing battlefield situational awareness,
2. Monitoring warfighter's health,
3. Remote training,
4. Real time fleet management,
5. Efficient inventory management.

Apart from the above-mentioned titles, of course, many more military fields can benefit from the Internet of Things. What should be considered here is whether the established model has the characteristics of the Internet of Things. To be more clear, information should be collected with the help of a sensor in military applications of the Internet of Things. The collected information should be transmitted at the network layer with the help of Internet and finally, this information is stored in the cloud storage. In addition, IoT applications should not be confused with cyber physics systems. As mentioned in the Internet of Things chapter, the decision-making mechanism in these models belongs to the decision maker. The IoT model has no decision-making responsibility.

Another issue that needs to be interpreted regarding the use of the Internet of Things in the military field is whether the Internet of Things applications should be fully autonomous or human-supervised. In IoT technology, objects are usually connected to each other via the Internet and the collected information is stored in a cloud system. The stored information is used by decision makers. This model can be established in the field of logistics and intelligence in the military. We can call this system autonomous systems. However, since warfare is a dynamic process and requires constant supervision of the decision maker, it is not possible for the command of the war to take place without human supervision in today's technology. So when it comes to command control of combat, as mentioned earlier in the Internet of Things chapter, the human being himself can be thought of as an object. Information received from soldiers or troops in the field can be used by decision makers. Fast and instant information from the battlefield can ensure the effective establishment of the decision support system. To sum up, human based information from the battlefield by decision makers reveals the necessity of human surveillance of the Internet of Things model established on the battlefield [19].

Applications of IoT in Military

In the military field, the Internet of Things can be used in many areas. However, in order to get an idea about the usage areas in general, it will be sufficient to give the following examples.

IoT-Based Smart Sniper

Bhomble et al. suggested a concept for real time observation and identification of adversaries using a smart sniper. A Webcam installed on the rifle gathers live footage from the actual conflict zone and communicates it to the control unit through wireless communication if the sniper spots an adversary. Once a target is spotted, the operator may position the sniper and fire at it using graphical user interface (GUI) buttons built in Python on the control unit [20].

Health Monitoring

It has become a great necessity in today's conditions to constantly monitor the health status of the soldiers in the battlefield. Because today, war can take place under very difficult conditions in various geographies. For example, whether a soldier has hypothermia can be checked by constantly monitoring body temperature values. Or, in a geography where an epidemic is common, the symptoms of the disease can be detected early, thanks to biochips, and the spread of the disease within the troops can be prevented. In this way, many people's lives can be saved or soldiers can be prevented from being out of war. Reyes Rolando et al. propose a new technology which is named as MilNova. This is an approach to the solution and research proposal for the implementation of the ubiquitous monitoring of military health in real time [21].

Security of Ammunition

For military units, the protection and control of ammunition is very important. Not being able to reach a needed ammunition when it is needed can change the course of the war. Or a change in the humidity values of the environment may cause the ammunition to be lost. Most military units already have various alarm systems for ammunition depots. But Liang Cao, Guo Zheng and Yong Shen developed a higher level ammunition control and tracking system. In this model, an ammunition container has become safer thanks to a ZigBee wireless sensor and various temperature, humidity, smoke and toxic gas sensors. In this way, it is ensured that the container is protected during shipment [22].

3 Conclusion

Thanks to the developments in information technologies, we can say that we are at the dawn of more frequent use of technologies in the military field such as IoT, Blockchain technology and its benefits about traceability and authorization. In our age information is needed more and more day by day and big data turns into fat data. The purpose of these applications is to enable the military command and control mechanism to make quick and secure decisions and to have the necessary information or asset when needed.

Also, we have learned that military operations could not be carried out and sustained without logistics however in today's world for business continuity a modern logistics is a must. Due to requirements in a complex battle environment experts must come up with solutions that use of the advanced technology like Industry 4.0 which is consist of supply chain management, progressive methods and processes.

References

1. NATO (2017) North Atlantic treaty organization: https://www.nato.int/cps/en/natohq/topics_61741.htm. adresinden alındı
2. Fernández-Villacañas Marín MA (2019) The transformation of the defense and security sector to the new logistics 4.0: public–private cooperation as a necessary catalyst strategy. In: Developments and advances in defense and security, smart innovation, systems and technologies, p 152
3. Benoit Montreuil RD (2012) Physical internet foundations. IFAC Proc Volumes 45(6):26–30
4. Supply Chain 4.0—the next-generation digitalsupply chain (2016) https://www.mckinsey.com/business-functions/operations/our-insights/supply-chain-40-the-next-generation-digital-supply-chain. adresinden alındı
5. Acero R, Torralba M, Pérez-Moya R, Pozo JA (2019) Order processing improvement in military logistics by value streamorder processing improvement in military logistics by value stream analysis lean methodology. In: 8th International conference on manufacturing engineering society, procedia manufacturing, vol 41, s 74–81

6. Wehberg GG (2020) Digital supply chains. In: Key facilitator to industry 4.0 and new business models, leveraging S/4 HANA and Beyond, 3rd edn. Taylor & Francis Ltd
7. Peter L (2020) Logistics in war military logistics and its impact on modern warfare. Logistics, digital transformation and future logisticians. https://logisticsinwar.com/2020/05/24/logistics-digital-transformation-and-future-logisticians/. adresinden alındı
8. Military Logistics (2018) This is an official U.S. Army: U.S. Army maneuver center of excellence: https://www.benning.army.mil/mssp/Logistics/. adresinden alındı
9. Wang Y (2006) Rand Corporation, Santa Monica, California, from factory to foxhole: improving the army's supply chain, in frontiers of engineering reports on leading-edge engineering. In: 2006 symposium national academy of engineering of the national academies The National Academies Press. National Academies Press, Washington
10. Peltz EH (2005) Sustainment of army forces in operation iraqi freedom: major findings and recommendations. MG-342-A. RAND Corporation, California, s 73
11. Solomon M (2020) Freight all kinds: making military logistics work in the 2020s and beyond. https://www.freightwaves.com/news/freight-all-kinds-making-military-logistics-work-in-the-2020s-and-beyond/amp. adresinden alındı
12. European Union Institute for Security Studies-EUISS (2017) https://www.iss.europa.eu/sites/default/files/EUISSFiles/Brief%2030%20Smart%20logistics.pdf. adresinden alındı
13. Joint Publications 4–01 The Defense Transportation System (2017) Joint chiefs of staff: https://www.jcs.mil/Portals/36/Documents/Doctrine/pubs/jp4_01_20170718.pdf. adresinden alındı
14. Hester JS (2009) A technique for determining viable military logistics support alternatives. School of Aerospace Engineering: Georgia Institute of Technology, pp 28–31
15. Maintenance Operations Headquarters (2019) Department of the Army: https://armypubs.army.mil/epubs/DR_pubs/DR_a/pdf/web/ARN19571_ATP%204-33%20C1%20FINAL%20WEB.pdf. adresinden alındı
16. Stevenson MH (2021) The Chief of Ordnance discusses a key facet of army transformation: the institution of a two-levelmaintenance system to replacetoday's four levels. In: Army maintenance transformation: https://alu.army.mil/alog/issues/SepOct02/MS838. adresinden alındı
17. Cusack J (2017) Standard chartered. Standard chartered: https://www.sc.com/en/explore-our-world/blockchain/. adresinden alındı
18. Rahayu SB (2019) Military blockchain for supply chain management. J Edu Soc Sci (JESOC) 13(1):353–361
19. Burmaoglu SS (2019) Defense 4.0: internet of things in military. In: Numerical methods for optimal control problems, pp 303–320
20. Bhomble BK (2017) IoT based smart sniper. In: 2017 international conference on I-SMAC (IoT in social, mobile, analytical and cloud) (I-SMAC)
21. Gotarane V (2019) IoT practices in military applications. In: 2019 3rd international conference on trends in electronics and informatics (ICOEI)
22. Cao LZ (2016) Research on design of military ammunition container monitoring system based on IoT. In: 2016 prognostics and system health management conference (PHM-Chengdu)

Toward Industry 5.0: Cognitive Cyber-Physical System

Zohreh Saadati and Reza Vatankhah Barenji

Abstract Industry 5.0 is regarded as the next industrial revolution; its objective is to combine the creativity of human experts with efficient, intelligent, and precise machines to produce manufacturing solutions that are resource-efficient and user-preferred. Cognitive cyber-physical technology is anticipated to enable Industry 5.0 in boosting productivity and delivering spontaneously customized products. This chapter provides an overview of cognitive cyber-physical systems, with an emphasis on the "cognitive" aspect of the technology.

1 Introduction

"Cognitive Science" is of, relating to, or involving conscious mental activities such as thinking, understanding, learning, and remembering. It commonly deals with the analysis, design, and development of social systems and human decision-making. On the other hand, "Intelligent system" is a system in which a machine is controlled or monitored by a cyber-physical system (CPS). CPSs are composed of several core components such as control systems, data analysis, IoT, information management systems, networking, and security. CPSs can be applied in a variety of fields including agriculture, education, clinical infrastructure, medical healthcare service, pharmaceutical industry, industrial automation, pharmaceutical industry, and transportation system. The main features of CPSs are pervasive computing, vast networks, reorganization, level of automation, and interactions with and without supervision. CPSs development demands three main steps as algorithm/architecture design, computation development, and implementation. In this chapter, we will discuss about

Z. Saadati (✉)
Smart Engineering and Health Research Group, Hacettepe University, Ankara, Turkey
e-mail: zohreh.saadati@outlook.com

R. V. Barenji
Department of Engineering, School of Science and Technology, Nottingham Trent University, Nottingham NG11 8NS, UK

"Cognitive-Cyber Physical System (C-CPS)" that might be the main enabler technology for next industrial revolution (i.e., Industry 5.0). The focus will be on the key cognitive-based functions of the C-CPS and its features and components.

2 Industrial Revolution

Throughout history, vehicles, clothes, housing, and weaponry have been developed and manufactured either by humans or with the aid of animals. In 1780, after the emergence of Industry 1.0, industrial production began to undergo substantial changes. In the first three revolutions, development process took decades to grow, while it took just 30 years to reach from the third to the fourth, and it is estimated that it will take less than 20 years to emerge into industry 5.0. In 1800s, Industry 1.0, machine-driven production infrastructures were developed for water and steam-powered equipment. As production capacity has increased, economic growth has improved substantially. Industry 2.0 began in 1900 with the introduction of electric power and assembly line manufacturing. Industry 2.0 was largely concerned with mass production, task allocation, and standards that boosted manufacturing organizations' productivity. Industry 3.0 emerged in 1970 with the advent of automation, electronics, information technology, and mass customization. Industry 4.0 has recently grown to include the notion of smart manufacturing. The goal is to create smart products utilizing cyber-physical system technology to maximize productivity. In the future, it is expected that Industry 5.0 will be incorporated by human expert creativity combined with efficient, intelligent, accurate machines empowered by CPS technology. The union of humans with CPS technology can be called as cognitive cyber-physical system (C-CPS).

Figure 1 shows an overview of the industrial revolutions since the eighteenth century to present which have shaped the political, ecological, and cultural spheres of Earth. Specifically, some technological advancements that led to the transition from the 3rd to the 4th industrial revolution and industry 5.0 by the introduction of AI into manufacturing. Overall technological advancement, since 1780, can be summarized as: miniaturization of integrated circuits, increased processing power, improved network bandwidth, improved mobile coverage, affordable microcomputers, widespread adoption of information technologies, and C-CPS.

2.1 Industry 5.0 Evolution Drivers

Industry 4.0 has revolutionized the production by merging a variety of technologies, including artificial intelligence (AI), the Internet of Things (IoT), cloud computing, and cyber-physical systems (CPSs). The main principle behind Industry 4.0 is to make production "smart" by integrating equipment and gadgets that can communicate with one another throughout their existence. Industry 4.0 places a premium on process

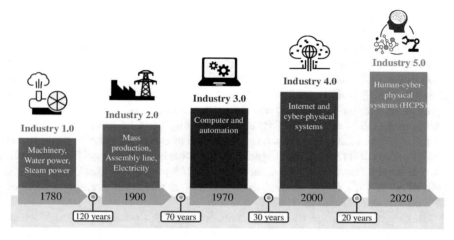

Fig. 1 Industrial revolution timeline

automation, which reduces human interaction in the production process. Industry 4.0 is focused on performance optimization through the use of CPS to connect equipment and applications.

Industry 5.0 may be seen as a way to combine the unmatched creativity of human expertise with powerful, intelligent, and precise machines. Industry 5.0 is supposed to restore the human touch to production. It will combine the capabilities of high speed, precise machines with the critical, cognitive thinking of people. Genius machines/systems and humans may be the primary outcome of Industry 5.0, in which humans can learn about machines/systems and machines can learn from humans. As a result of this mix, very intelligent goods will be created.

Industry 5.0 will greatly improve production productivity, quality, and speed to market, while also enabling human–machine interaction and ongoing monitoring. The partnership of people and robots is aimed at rapidly expanding output. Industry 5.0 will be able to improve production quality by delegating repetitive and boring activities to robots and critical thinking tasks to people. In comparison to Industry 4.0, Industry 5.0 would encourage more skilled occupations due to intellectual experts' interaction with machines. Another intriguing benefit of Industry 5.0 will be the availability of more environmentally friendly solutions in comparison to contemporary industrial transformations, neither of which places a premium on environmental protection.

3 Cyber-Physical System

Commonly, autonomous machines do not require a person to accomplish difficult tasks or jobs. Instead, they combine human skills to accomplish these tasks.

CPS is the main enabler technology to develop autonomous machines and industry advancements. By enhancing system interaction and making the smart machines, CPS surpasses traditional computing systems such as PLC in terms of computational power, robustness, and synergy in autonomous machines/systems. Around 2006, Helen Gill coined the term CPSs while working at the National Science Foundation (NSF) [1]. In 2007, CPS became a special track at the IEEE Real-Time Systems Symposium and, the same year, the President's Council of Advisors on Science and Technology (USA) announced CPS as a top priority [2]. CPS week was organized for the first time in 2008, and it is now recognized as one of the premier conferences on CPS. The conference brings together the top five conferences on CPS research. The National Science Foundation launched the CPS program in 2009, and in 2010, both ACM and IEEE cosponsored the CPS Conference. After that, the number of CPS conferences increased rapidly.

On October 2013, the European Commission held a workshop titled "Cyber-Physical Systems: Uplifting Europe's Innovation Capacity" to identify industrial, scientific, technological, and socioeconomic challenges, European strengths, stakeholders, and innovation priorities in the field of CPS. Germany's Industry 4.0 initiative is a direct result of the US CPS initiative. In the following years, a larger number of coordination and support action projects were launched in Europe, including CyPhERS [3], CPSoS [4], Platforms4CPS [5], and others. For instance, the CPSoS project—toward a European roadmap on research and innovation in engineering and management of cyber-physical systems of systems—was designed to create constituencies for a European research and innovation agenda on systems-of-systems. CPSoS is a forum and exchange platform for communities and ongoing projects which focus on the challenges posed by the engineering and the operation of technical systems that interact with large complex physical systems. These studies considered CPS as the key enabling and disruptive technology required for securing economic leadership in embedded systems and information and communication technologies, with significant social and economic impact [6]. CPSE-Labs [7], one of the most recent initiatives in CPS research, found that CPS can contribute decisively to societal challenges such as coping with an aging population, addressing climate change, improving health and safety issues, and supporting the switch to renewable energy. A cyber-physical system is a system in which computation is integrated with physical processes, where embedded computers and networks monitor and control those processes, often in feedback loops where the physical processes influence computations and vice versa. In other words, CPS is defined as the integration of analog, digital, physical, and human systems using logic and physical systems. It provides infrastructure to enable smart services in real-world settings.

4   Applications of Cognitive Cyber-Physical System

Cyber-physical systems (CPS), which involve physical and engineered systems together with computing and networking, are attracting extensive attention in areas as diverse as smart grid, environment monitoring, vehicular networks, and industrial automation, driven by the growth in Internet of Things, wearable devices, cloud computing, and big data analysis. Cognitive methods can be used in many novel applications of CPS to improve the system's performance [8]. The Cognitive Cyber-Physical System is applied into various areas such as agriculture, education, industrial automation, medical healthcare device, pharmaceutical industry, smart manufacturing, security, smart city and home, transportation systems, etc. The following section described some of Cognitive Cyber-Physical Systems applications and working procedure as below.

5 Cognitive Cyber-Physical Systems (C-CPS)

Human–machine collaboration (HMC) is the integration of human activity with machine processes. HMC is used in a variety of fields, including surgeons, assemble line, manufacturing, elder care, and rehabilitation. HMC can be used to process household and sorting tasks or for using physical contact between machines and humans to provide information. Integrating the CPS and cognitive science technologies is a hallmark of HMC. The CPS provides decision and control functions, while cognitive science provides the how the human mind works. C-CPS refers to the integration of cognitive radios into the CPS, which provides the cognitive capability to obtain information regarding transmission requirements or spectrum availability to enhance the performance of the system. C-CPS empowers human and machines interoperability to solve everyday tasks and interaction with the real world. The C-CPS consists of a control system, task monitoring algorithms, human-side control algorithms, and task command algorithms. As part of the C-CPS, there is the human integration component, the human–machine collaboration component, as well as the machine operating system or real-time control operating system. Different collaboration protocols may be used when combining the CPS and cognitive science technologies [9].

Over the past few years, C-CPS has drawn significant attention from academia and industry. Such interest can be attributed to the potential of these technologies to revolutionize human life by combining robust computation with complex visual scenes, in which environmental conditions may change, adapting to a wide range of unforeseen changes, and predicting possible behaviors based on cognitive capabilities that are able to sense, analyze, and act accordingly. Nevertheless, due to the complexity of such processes, perception of the environment and converting it to the knowledge relevant for the decision-making remains as a challenge for real-time applications.

C-CPS consists of different components. The components differ according to the application. Mechatronics, cybernetics, process, and design are all ideas incorporated into the physical system. Embedded or built-in systems are frequently used to refer to the control of physical systems. Embedded systems are frequently constructed in such a way that contact between their physical and computer components is minimized or limited. The architecture and communication of the C-CPS are likewise comparable to those of the Internet of Things (IoT) since they employ the same fundamental IoT concepts. Nonetheless, CPS facilitates the interaction of physical and computational components at a high level. For example, intelligent factories integrate a range of components. All the components, including portable devices, connecting devices, RFID readers, RFID tags, and online and cloud storage, are connected.

The C-CPS contains nodes capable of sensing, analyzing, and reacting to the environment in accordance with their analysis. Also, it includes cyber-physical element that facilitates interconnectivity among all elements. The challenge for such an architecture is to understand and model the human cognition, which includes decision-making, perception, learning, and adaptivity, as well as control, learning, and adaptivity. For cognitive cyber-physical systems, knowledge and learning are key components for decision-making. Knowledge must be explicitly and systematically represented so that it can be articulated, codified, accessible, and shared across interconnected elements and platforms. Knowledge itself is not a trivial task, given the many sources of information needed to generate and represent this knowledge. Likewise, human intervention or autonomous learning is the only way to improve decision-making. Experience in both cases is taken into consideration. Thus, in order to build an advanced C-CPS, past experience must also be represented. Cognitive CPS's can learn from each other, from humans, and can also form collaborative networks [10].

The following section presents cognitive cyber-physical system, key cognitive functions in C-CPS, and some potential applications of C-CPS.

6 Key Cognitive Functions in C-CPS

Cognition is the process of transforming, reducing, elaborating, storing, retrieving, and using sensory input [11]. The study of cognition is known as cognitive psychology. It involves inner mental processes such as attention, perception, memory, problem-solving, and knowledge representation (Fig. 2).

Our identity and behavior are shaped by each of these components. C-CPS is derived from advances in cognitive science, machine learning, and artificial intelligence and achieve some critical elements of cognitive function including attention, perception, memory, reasoning, and problem-solving. As a result of C-CPS technology, enterprises can exploit implicit knowledge creatively, effectively, and efficiently, enabling the transfer of higher performance decisions and controls as well as improving performance across the enterprise. The following sections explain each cognitive component in C-CPS as below:

Fig. 2 Levels of cognition

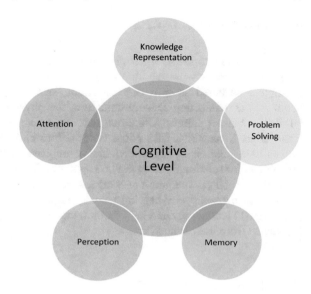

6.1 Attention in C-CPS

Attention is selective focusing on the most relevant information available from multiple sources. Neurocomputational models demonstrate the importance of attention in cognition. Human, robot, and driving vision have been major focus areas for modeling. Some of the theories developed to figure out at what point we are able to notice, pay attention, and identify objects like feature integration theory, guided search model, CODE theory of visual attention, signal detection theory, and computational models like Itti's model: color, intensity, orientation, decision theory, information theory, graphic models, and spectrum analysis model. Attention is divided into two levels. The first level of attention is focused on a specific task. Top-down searching for relevant information is the second level. C-CPS requires attention to changes in conditions and contexts, resilience and fault detection, as well as hierarchical and programmable attention models. Here are some examples of programmable attention models:

1. Deep learning models based on self-attention
2. Image recognition using non-local neural networks
3. An attentive meta-learner.

According to Khargonekar [10] some of the related work in C-CPS model of attention are as Recurrent Models of Visual Attention, Effective Approaches to Attention-based Neural Machine, Show, Attend, and Tell, Neural Image Caption Generation with Visual Attention, and Self-attention Generic Adversarial Networks (GANs).

6.2   Perception in C-CPS

Perception refers to the interpretation of vision, sounds, smells, and touches. In C-CPS, perception is representation of data in an effective way; it involves sensing multiple distributed sources of information, and it is a process of interpreting, acquiring, and selecting sensory data from the physical world to develop actions similar to human actions. Cognitive theories of perception heavily rely on prior knowledge. Sensor-rich CPS could benefit from combining neural network techniques with relational prior knowledge to improve context awareness. Tools and techniques for relational priors include symbolic front ends with priors to learn the symbolic front ends and graph networks, such as scene graphs. For example, virtual Warsaw, it is a smart city helping visually impaired residents based on Internet of Things (IoT) technology and help them to move independently around the city with assistance from their smartphones.

6.3   Memory and C-CPS

A memory is a way of encoding and retrieving information over time. Memory is essential to intelligent behavior. Memory mechanisms in human cognition include short term, long term, episodic (content addressable), and semantic. An excellent example of how memory is used in machine learning is the long short-term memory (LSTM). An LSTM layer consists of a set of recurrently connected blocks, known as memory blocks. An analogy to these blocks is that they are differentiable versions of the memory chips in digital computers. There are one or more recurrently connected memory cells as well as three multiplicative units—input, output, and forget gates—that provide continuous analogs of write, read, and reset operations. The LSTM network is a type of recurrent neural network capable of learning-order dependence in sequence prediction problems. A behavior like this is required in complex problem domains like machine translation and speech recognition. Differential neural computers (DNCs) by Graves et al. are neural networks which take advantage of both memory augmentation and attention mechanisms. DNC has the capability of learning aspects of symbolic reasoning and applying them to working memory. Experience replay is an innovation in Deep RL. Experience replay is a replay memory technique used in reinforcement learning, whereby we accumulate experiences at each time-step, $e_t = s_t, a_t, r_t, s_t + 1$, into a data-set $D = e1,...,eN$, which is then combined over many episodes into a replay memory set. Additionally, sparse distributed representations. Examples are hierarchical temporary memory and sparsey. The memory of CPS is used to reenact successful episodes. Small data learning and control can be influenced by episodic control. Recorded experiences are used as value function estimators in model-free episodic control. And using deep learning models and lookup tables of action values, neural episodic control is achieved.

6.4 Problem-Solving in C-CPS

The process of discovering, analyzing, and solving a problem involves the mental process of problem-solving. To make optimal decisions, C-CPS must be able to solve problems. In the context of goal-oriented tasks, reinforcement learning ideas and techniques are useful. Effective and scalable reinforcement learning (RL) algorithms are required. Safety is of utmost importance in CPS. There are several possible approaches to RL, including model-based RL, safe RL algorithms, and hierarchical RL.

The levels of maturity for CPS are as follows: Setting basics, creating transparency, enhancing understanding, improving decision-making, and, finally, self-optimizing. At the first level, the organizational and structural conditions for implementing CPS are created, while at the four higher levels, the realizations about information and knowledge processing, and aspects of cooperation and collaboration are mature. "Information generation" refers to the requirement for real-time data availability for all CPS activities, while "Information processing" refers to the existing tools used to deduce new knowledge. Achieving the most sophisticated layer, "Information linking" refers to collaboration-based adaptations of CPS processes, whereas "interacting cyber-physical systems" is the most complex layer that can only be achieved by independent problem-solving capabilities of collaborative CPS [12].

6.5 Knowledge Representation and Reasoning in C-CPS

Knowledge representation and reasoning (KR and R) is a field of artificial intelligence that involves the development of intelligent systems that understand their environment and are able to act upon it automatically, much like humans. By KR function, a computer uses it to represent information about the real world, so it can understand it and use it to solve complex real-world problems such as diagnosing a medical condition or communicating with people. Also, it describes how knowledge can be represented in artificial intelligence. In addition to storing data into some database, knowledge representation enables intelligent machines to learn from that data and experiences, so they can behave intelligently like humans. In KR and R, a fundamental assumption is that knowledge is represented as concepts, logic, rules, procedures, and ontologies for understanding relationships between entities. However, ontologies alone cannot represent knowledge since they aim to represent relations between different entities. Therefore, a well-defined data model, which identifies the entities, their relationships, constraints on these entities, and their data types, is essential for monitoring [13].

In smart manufacturing, knowledge representation and reasoning play an important role in decision-making at runtime. Modern automatic control systems should be able to integrate seamlessly and assist operators. CPS uses semantic models as knowledge repositories, and service orchestrators use ontologies to organize and schedule execution processes using all available resources, including human operation. With this approach, the operator acts as an active asset for the orchestration engine to consider in production plans.

7 Cognitive Cyber-Physical System Features

The following are some of the main characteristics of C-CPS systems:

- In C-CPS, each component has cognitive abilities such as perception, decision-making, communication, and collaboration. C-CPS can each make production-related decisions independently;
- Enhanced the industry's current intelligence and independence by improving key components and systems;
- C-CPS is capable of adapting to production changes in a very short period;
- It is essential to provide portability, local storage, and central storage;
- A system that integrates multiple discrete devices.

8 Discussion

In comparison to Industry 4.0, Industry 5.0 aims to leverage human creativity to generate more resource-efficient and user-friendly manufacturing solutions by integrating human innovation with efficient, intelligent, and precise machinery. Industry 5.0, it is commonly assumed, will reintroduce the human touch to production. Industry 5.0 is projected to combine great speed and precision with critical thinking and cognitive ability. Industry 5.0 also advances mass customization, allowing clients to select customized, and customized items based on their requirements and tastes. Industry 5.0 will boost industrial efficiency and adaptability between humans and machines, allowing for continuous monitoring and interaction. Humans and robots collaborate to rapidly boost production. To improve the quality of production in Industry 5.0, repetitive and monotonous jobs can be delegated to robots and machines, while critical thinking duties are delegated to people. Humans will guide robots via mass customization in Industry 5.0. While Industry 4.0 is focused on connecting CPSs, Industry 5.0 connects Industry 4.0 applications to collaborative robots (cobots). Industry 5.0's objective is to create more accurate and stable judgments via the use of predictive analytics and operational intelligence. The majority of manufacturing processes in Industry 5.0 will be automated since real-time data from machines will be combined with highly qualified individuals.

Numerous industry practitioners and scholars have offered a variety of definitions for Industry 5.0, which is still in its infancy. Here are a few definitions. Industry 5.0 reintroduces the human workforce to the industry through the integration of workflows and intelligent systems, where human and machine collaborate to boost process efficiency. Industry 5.0 provides individualized autonomous manufacturing through the use of the ideal human companion and cobots. Thus, the person and the machine are able to collaborate. While cobots lack programmed skills, they are capable of sensing and comprehending the presence of people. Cobots will do repetitive chores and manual labor, while human employees will handle customization and critical thinking.

9 Conclusion

Cognitive cyber-physical systems widely will be utilized in many applications for a variety of purposes. C-CPS will be applied in analyses, designs, decision-making, and socio-technical systems in various areas such as agriculture, education, medical healthcare service, pharmaceutical industry, industrial automation, manufacturing, and transportation service. Each application relies on human interaction and group interaction. Control system, networking, Internet of Things (IoT), data analysis, security, and information management system are all critical components of a cognitive system. The characteristics of cognitive cyber-physical systems include ubiquitous computing, huge networks, rearrangement, a high degree of automation, and interaction with and without supervision. Three layers of implementation are possible for a cognitive cyber-physical system: computational model, algorithm, and implementation. Each physical and cyber-component application in C-CPS has unique problems and design issues. These difficulties differ according to application. Each layer, abstraction, and architecture impose a unique set of design constraints. Numerous obstacles exist in terms of design and operation, including energy management, defect detection, resource management, and human interaction. One of the primary concerns in real-world situations is the expense of implementing CPS. Future work will focus on developing a system that can function in a variety of modes for a variety of applications and doing a cost analysis.

References

1. Lee EA, Seshia SA (2016) Introduction to embedded systems. A cyber-physical systems approach. MIT Press. https://doi.org/10.1007/71006
2. CyPhERS (2014a) CPS20: CPS 20 years from now
3. CyPhERS (2014b) Structuring of CPS Domain: Characteristics, trends, challenges and opportunities associated with CPS - Deliverable D2.2. CyPhERS - Cyber-Physical European Roadmap & Strategy, 611430, 41
4. CPSoS (2016) CPSoS – towards a european roadmap on research and innovation in engineering and management of cyber-physical systems of systems
5. Thompson H, Reimann M, Authors L, Ramos-hernandez D, Bageritz, S, Brunet A, Robinson C, Sautter B, Linzbach J, Pfeifer H, Aravantinos V (2018) Platforms4CPS Key Outcomes and Recommendations
6. Geisberger E (2015) Living in a networked world Integrated research agenda Cyber-Physical Systems (agendaCPS) (Issue March)

7. CPSE-Labs (2018) Why are Cyber-Physical Systems important? http://www.cpse-labs.eu
8. Tang K, Shi R, Shi H, Bhuiyan MZA, Luo E (2018) Secure beamforming for cognitive cyber-physical systems based on cognitive radio with wireless energy harvesting. Ad Hoc Netw 81:174–182
9. Mizanoor Rahman SM (2019) Cognitive cyber-physical system (C-CPS) for human-robot collaborative manufacturing. In: 2019 14th Annual conference system of systems engineering (SoSE), pp 125–130
10. Khargonekar PP (2020) Cognitive cyber-physical systems: cognitive neuroscience, machine learning, and control. In: Proceedings American Control Conference, pp 4757–4758
11. Neisser U (1994) Multiple systems: a new approach to cognitive theory. Eur J Cogn Psychol 6(3):225–241
12. Monostori L et al (2016) Cyber-physical systems in manufacturing. CIRP Ann Manuf Technol 65(2):621–641
13. Gürdür D et al (2018) Knowledge representation of cyber-physical systems for monitoring purpose. Proc CIRP 72:468–473
14. John A, Mohan S, Vianny DMM (2021) Cognitive cyber-physical system applications. Cogn Eng Next Gener Comput 167–187
15. Lee EA (2008) Cyber physical systems: design challenges. In: 2008 11th IEEE international symposium on object and component-oriented real-time distributed computing (ISORC)
16. Zhang Q, Liu Y (2022) Reliability evaluation of markov cyber–physical system oriented to cognition of equipment operating status. Comput Commun 181:80–89
17. Wiedermann J, van Leeuwen J (2021) Towards minimally conscious cyber-physical systems: a manifesto. In: SOFSEM 2021: theory and practice of computer science. Springer International Publishing, Cham, pp 43–55
18. de Oliveira CS, Sanin C, Szczerbicki E (2019) Visual content representation and retrieval for cognitive cyber physical systems. Proc Comput Sci 159:2249–2257
19. Szabo G, Racz S, Reider N, Munz HA, Peto J (2019) Digital twin: network provisioning of mission critical communication in cyber physical production systems. In: 2019 IEEE international conference on industry 4.0, artificial intelligence, and communications technology (IAICT)
20. Peebles D, Cooper RP (2015) Thirty years after Marr's vision: levels of analysis in cognitive science. Top Cogn Sci 7(2):187–190
21. Zhang J (2019) Cognitive functions of the brain: perception, attention and memory. arXiv [q-bio.NC]
22. Maddikunta PKR et al (2021) Industry 5.0: a survey on enabling technologies and potential applications. J Ind Inf Integr 100257
23. Shkodyrev VP (2016) Technical systems control: from mechatronics to cyber-physical systems. In: Studies in systems, decision and control. Springer International Publishing, Cham, pp 3–6
24. Wilhelm Frederik van der Vegte RWV (2013) Considering cognitive aspects in designing cyber-physical systems: an emerging need for transdisciplinarity. In: Proceedings of the international workshop on the future of transdisciplinary design TFTD, pp 41–52
25. Besold TR et al (2017) Neural-symbolic learning and reasoning: a survey and interpretation. arXiv [cs.AI]
26. Radanliev P, De Roure D, Van Kleek M, Santos O, Ani U (2020) Artificial intelligence in cyber physical systems. AI Soc 36(3):1–14
27. Demir MO, Topal OA, Pusane AE, Dartmann G, Ascheid G, Kurt GK (2021) An adaptive multi-agent physical layer security framework for cognitive cyber-physical systems. arXiv [eess.SY]
28. Şahinel D, Akpolat C, Görür OC, Sivrikaya F, Albayrak S (2021) Human modeling and interaction in cyber-physical systems: a reference framework. J Manuf Syst 59:367–385

Printed in the United States
by Baker & Taylor Publisher Services